Blockchain and AI

In the rapidly evolving landscape of the digital age, two technologies stand out for their transformative potential: Artificial Intelligence (AI) and Blockchain. This book offers an incisive exploration of the confluence between these technological titans, shedding light on the synergies, challenges, and innovations that arise at this intersection. The chapters explore thought-provoking analyses, informed by cutting-edge research and expert perspectives, that navigate the nuanced interplay of decentralized ledger technology and intelligent systems. From potential applications in teaching and learning, finance, healthcare, and governance to ethical considerations and future trajectories, this volume serves as an essential compendium for scholars, professionals, and anyone keen to grasp the future of digital innovation.

T0351008

Smart Technology: Methods, Contexts and Consequences
Series Editor: Niaz Chowdhury

Blockchain and AI: The Intersection of Trust and Intelligence
Editor: Niaz Chowdhury

For more information about the series, please visit: https://www.routledge.com/Smart-Technology/book-series/SMTECH

Blockchain and AI
The Intersection of Trust and Intelligence

Edited by
Niaz Chowdhury
Ganesh Chandra Deka

CRC Press
Taylor & Francis Group
Boca Raton London New York

CRC Press is an imprint of the
Taylor & Francis Group, an **informa** business

Designed cover image: © Shutterstock

First edition published 2024
by CRC Press
2385 NW Executive Center Drive, Suite 320, Boca Raton FL 33431

and by CRC Press
4 Park Square, Milton Park, Abingdon, Oxon, OX14 4RN

CRC Press is an imprint of Taylor & Francis Group, LLC

ISBN: 978-0-367-75327-6 (hbk)
ISBN: 978-0-367-75331-3 (pbk)
ISBN: 978-1-003-16201-8 (ebk)

DOI: 10.1201/9781003162018

Typeset in Caslon
by MPS Limited, Dehradun

Contents

Editor Biographies

Dr Niaz Chowdhury is a scientist, educator, start-up co-founder, author, and grant winner from the United Kingdom working with technology-aided 21st-century education and disruptive technologies, including Blockchain and Artificial Intelligence. His experience spans over one and a half decades across three British and Irish nations in England, Scotland, and Ireland.

Dr Chowdhury has co-founded two start-ups in the UK. One of them is the London Tech Lab, a deep tech research lab established in 2020 during the pandemic. The lab focuses on developing technologies related to Web3, AI, Internet of Things, Cybersecurity, and Fintech. Dr Chowdhury currently serves as the Chief Scientist at the lab. The other start-up is the London School of Leadership, Investment, and Technology (LSLIT), which was founded the following year. LSLIT is an EdTech institution providing executive education for professionals and lifelong learners. Dr Chowdhury holds the positions of Chairman and Director of Education at LSLIT.

Previously, he held the position of AI Lead in the UK Government's InnovateUK co-funded project SAM AI, where he was the P.I. winning the grant with Store Performance, the participating SME. He was also a researcher at Knowledge Media Institute (KMI), The Open University, UK, working with EU Horizon2020 projects QualiChain and DEL4ALL and Smart City project MK-Smart, collaborating with the University of Cambridge. During his time at the OU, he was a key team member that developed the world's first digital COVID-19 vaccination certification system.

He authored *Inside Blockchain, Bitcoin, and Cryptocurrencies* in 2019, published by Taylor & Francis, and edited two other books. He also published numerous research articles in reputed journals and conferences.

Dr Chowdhury earned his doctoral degree from the University of Glasgow, Scotland and was a Govt. of Ireland Scholar at Trinity College, University of Dublin, Ireland. He was also a Gold Medallist awarded by H.E. President of the People's Republic of Bangladesh.

Mr Ganesh Chandra Deka is Joint Director, RDSDE Assam, DGT in the Ministry of Skill Development and Entrepreneurship, Government of India.

His research interests include Blockchain Technology, BigData Analytics, NoSQL Database and Internet of Things (IoT). He has authored two books on cloud computing (Publisher-LAP Lambert, Germany), and co-authored five textbooks (four Fundamentals of Computer Science, one Free and Open Source Software).

Mr Deka has edited 29 books (Seven IGI Global, USA; Eight CRC Press, USA; Four Springer; and Ten Elsevier) on Bigdata, NoSQL Database and Blockchain Technology. He authored 12 book chapters, publishedeight papers in various reputed Journals (two IEEE, one Elsevier and five others) and around 47 research papers in various IEEE conferences. He was Guest Editor for four Special Issues of indexed International Journals (SCOPUS & SCI Journal).

As of now, he has organized eight IEEE International Conference as Technical Chair in India. He is an Editorial board member and reviewer for various journals and international conferences, member of IEEE, the Institution of Electronics and Telecommunication Engineers, India and Associate Member, the Institution of Engineers, India.

Contributors

Maruf Farhan is not only very knowledgeable about cybersecurity but also quite enthusiastic about it. Maruf has demonstrated serious commitment to his future by pursuing a Master of Science in cyber security at Northumbria University. His drive to further his studies shows his interest in the ever-changing cybersecurity industry.

Maruf was employed at Kaspersky Lab Technical Support in Malaysia from 2016 until 2020 in the capacity of Security Support Specialist. By working on real-world security issues, he was able to hone his expertise in the field. Soon after, he began working as an IT officer for a local IT firm. Maruf has been working in the IT industry for seven years, and he has earned various certifications from CompTIA for his efforts, including Security+ and A+, python and C+, and blockchain technology (Ethereum blockchain). These certifications attest to his proficiency in the very important field of IT security.

Besides his professional accomplishments, Maruf has contributed to academia. He has published in two international, peer-reviewed journals, and his book chapter will be his first. He practises what he preaches in cybersecurity. Maruf wants an academic career. Blockchain and anti-phishing are his research interests, and he wants a cybersecurity doctorate. He is forward-thinking, showing his interest in blockchain technology, a cybersecurity revolution, and phishing, a common internet threat.

All in all, Maruf Farhan is a driven young man with a strong foundation in cybersecurity, a commitment to lifelong learning, and lofty goals for his academic and professional future in the IT industry. He is a great choice because of his dedication to the field, which will help to further the advancement of cybersecurity practices.

Nazmul Hoque, Blockchain Intern at *London Tech Lab, UK.*

Md Aminul Islam is an engineer, teacher, and researcher. He studied several domains, including business, social science, education, and computer science. He holds a BSc and MSc in Computer Science and is currently doing research in AI. He has certification in education and training, networking, blockchain, and cloud and wrote eight books for college students in Bangla. He won a few gold awards in leadership, research, and extracurricular activities. He has a membership of IEEE, the British Computer Society, the Royal Statistical Society, and STEMResearchAI. As a philanthropist through charities like Rotary International holding leadership positions from club president, DRR and MDIO Secretary is continuing to contribute to the development of the community. His focus of research is AI, ML, and Edtech.

Reza Nourmohammadi is a Post-Doctoral Fellow at the UBC Blockchain Lab. He has also completed a PhD in the area of blockchain and machine learning integration. It is proposed to design an adaptive public blockchain that is able to learn and adjust itself as network conditions change. In addition to his BS degree in Computer Engineering, Reza holds a Master's degree in Artificial Intelligence. He had a background in cryptography during his academic years.

In addition, he spent a considerable amount of time working in the financial industry. He was inspired by this to become involved in blockchain technology. Since 2015, he has been involved in the blockchain industry both in theory and practice, and he has worked as a professional in the IT industry since 2003.

Over the past three years, he has volunteered as a blockchain lecturer and taught people about this fascinating and innovative technology and currently, he is engaged in developing a decentralized artificial intelligence (AI) system based on blockchain technology in order to maximize its impact on everyday life.

Naahi Mumtaj Rihan is studying Electrical & Electronics Engineering at BRAC University of Bangladesh. She is passionate about creating a discrimination-free sustainable city and a world with less environmental pollution. Rihan plans to research Aerospace Engineering and work in sustainable aerospace engineering. Apart from these, she is a robotics hobbyist, sustainable, and green energy enthusiast, climate activist, thalassophile, and tech enthusiast along with having a passion for marine conversation and marine robotics.

Md Abu Sufian, based in Birmingham, UK, is a distinguished researcher in Artificial Intelligence, Machine Learning, and Data Science. His prolific academic portfolio includes publications in conference proceedings, journals, and book chapters. Some of his seminal works explore the intersection of AI in sectors like MedTech, Healthcare and the UK Energy Sector. Sufian has also graced various international conferences from London, Paris and California, presenting his pioneering research. His couple of research papers are ready to be presented at different international conferences. Holding an MSc in Data Analysis for Business Intelligence from the University of Leicester, he furthered his expertise through an internship with ASAP Data Solution Ltd and impactful projects with HSBC-Financial Services in London. Skilled in tools such as Python, R Studio, Azure and Power BI, Abu Sufian blends technical acumen with a profound understanding of data to influence strategic decision-making. His contributions stand as a beacon in the realm of AI and data-driven research.

Preface

In the sprawling landscape of the 21st century's digital revolution, two technologies have emerged as beacons of transformative potential: Blockchain Technology and Artificial Intelligence (AI). On the one hand, Blockchain, with its transparent and immutable ledger system, promises an unprecedented level of trust and security in digital transactions. On the other hand, AI, with its ability to process and analyse vast amounts of data, offers the promise of unlocking unparalleled efficiencies and insights. Individually, these technologies are powerful; together, their potential is magnified manifold, hinting at a synergy that could revolutionize multiple industries.

The aim of this book is to explore this synergistic relationship, diving deep into the intricacies of how Blockchain and AI can come together to provide solutions to age-old problems and modern challenges alike. From the medical field, where patient data management and diagnosis can achieve new levels of accuracy and security, to the world of finance, where transactions can be both ultra-efficient and transparent, the applications are vast and varied.

However, the potential of Blockchain and AI is not just limited to industries with monetary undertones. Consider education, for instance. These technologies could offer solutions for secure record-keeping, transparent credential verification, and personalized learning experiences tailored to individual student needs. With AI analysing student data and Blockchain ensuring the integrity and accessibility of educational records, the educational landscape could be on the brink of a groundbreaking transformation.

The inspiration for this book was drawn from a simple observation: while there are numerous texts examining Blockchain and AI independently, few venture into the vast domain where these technologies intersect. This is surprising given that the overlap could be where the most transformative potentials lie. It's at this intersection that trust meets intelligence, security meets prediction, and transparency meets analysis.

Each chapter in this book is carefully crafted to provide insights, not just from a technical standpoint but also from a holistic, industry-specific perspective. Real-world examples and case studies are intertwined with theoretical knowledge to offer a comprehensive understanding. Whether you're an industry expert, a tech enthusiast, or a curious reader, this book aims to enlighten, inform, and inspire.

However, it is crucial to remember that like all technologies, Blockchain and AI are tools. Their efficacy and ethicality lie in their applications. As we stand on the brink of what might be another industrial revolution, there is an inherent responsibility to harness these tools judiciously. This book also explores the challenges, limitations, and ethical implications of integrating Blockchain and AI, underscoring the importance of responsible innovation.

Last but not least, the fusion of Blockchain and AI is more than just about technological advancement; it is about crafting a future that is transparent, efficient, and inclusive. Through the pages of this book, we invite you to embark on a journey to explore this promising frontier, hopeful that it sparks both understanding and imagination.

Editors
Niaz Chowdhury
Ganesh Chandra Deka

1

BLOCKCHAIN EMPOWERED FEDERATED LEARNING

REZA NOURMOHAMMADI

1.1 Introduction

Over the past years, the field of Artificial Intelligence (AI) has piqued increasing interest, permeating various domains such as driverless cars, medical care, and finance. AI has also become an integral part of people's daily lives through applications like recommender systems on social media and streaming services. In this chapter, our focus will be on one specific aspect of AI: machine learning (ML).

As highlighted by Yang et al. [1], despite the substantial efforts made by researchers and industries to enhance ML algorithms, the true impediment to ML's success lies in the scarcity of user data.

In the big data era, one of the most critical intelligent applications is multiparty learning or federated learning. As institutions, companies, and smart devices collect vast amounts of data daily, the conventional approach of centralizing all data for learning becomes inefficient and insecure. In multiparty learning, the process is distributed, enabling each party to keep their data locally. Most existing multiparty learning systems focus on training a global model for all parties, neglecting the potential of utilizing local models already trained on individual datasets. Effectively utilizing these local models can substantially improve the training efficiency of multiparty learning. Some researchers have explored model averaging techniques to enhance performance and proposed methods resilient to Byzantine attacks.

However, a challenge arises from assuming that local models at each party are homogeneous, which may not be practical. Some works attempt to address this by proposing methods to calibrate heterogeneous local models in the system. Yet, these approaches often rely on a

DOI: 10.1201/9781003162018-1

trusted central server to coordinate the distributed learning process, introducing a single point of failure and susceptibility to attacks. Only a few works have ventured into designing decentralized multiparty learning systems, with some focusing solely on linear models.

Traditional ML approaches involve centralized data collection and training, raising significant challenges concerning data sensitivity, particularly with respect to privacy. For instance, in medical research involving multiple centers, sharing sensitive patient information for model training may not be feasible due to privacy concerns. The same privacy issues apply to personal smartphones with vast amounts of data, such as search preferences, when centralized in data centers.

To address these challenges, distributed ML has emerged as a potential solution. Leveraging the increasing computing power of user devices, distributed ML allows training models locally while exchanging only necessary information, preserving data privacy and security. This approach represents a step toward overcoming the limitations of traditional centralized learning methods and presents an open and challenging problem for secure decentralized multiparty learning with heterogeneous models, requiring further research and innovative solutions.

Federated machine learning (FML) operates by forming a Federation of clients that compute local model updates (i.e., training) based on their respective datasets. These updates are then averaged by a Central Coordinator Server to create a global model. This process is repeated for a certain number of communication rounds, and the algorithm used is called Federated Averaging (FedAvg). Thanks to FML, the benefits of obtaining a global model are retained, and participants' data privacy is preserved as sensitive information is not disclosed.

Algorithm 1.1: Federated Averaging
```
# Initialization
Initialize global model parameters θ
Initialize empty list of client datasets D = {D_1, D_2, ... , D_N}
```

Set hyper-parameters: learning_rate, num_rounds, num_clients_per_round, batch_size

```
# Training loop
for round = 1 to num_rounds:
    selected_clients = Randomly select num_clients_per_round clients from D
    aggregated_gradients = 0

    for each client in selected_clients:
        # Train local model on client's data
        local_model = Initialize a copy of the global model with parameters θ
        local_data = Sample batch_size data points from client's dataset
        for each data point in local_data:
            Compute gradient of loss with respect to local_model parameters
            aggregated_gradients += Computed gradient

    # Average gradients across clients
    averaged_gradients = aggregated_gradients/num_clients_per_round

    # Update global model
    θ = θ - learning_rate * averaged_gradients

# Final global model after training
Final_global_model = θ
```

In summary, FML presents a promising approach to overcoming the challenges of data privacy while still reaping the advantages of collective intelligence in the field of ML.

1.2 Federated Learning

The proliferation of distributed data streams generated by Internet of Things (IoT) devices has significantly contributed to the development of powerful and precise ML models. These models find applications in various industries, including cutting-edge technologies such as Industry 4.0, cyber-physical systems, and smart mobility [2]. As a result, the processing and utilization of these data streams have become highly valuable, leading to a significant surge in their usage.

However, this increased reliance on distributed data comes with a challenge—much of these data either originate directly from human users or contain sensitive information about human behavior, raising concerns about privacy when transmitting edge data on a large scale.

Furthermore, companies are increasingly hesitant to share their data due to the growing awareness of privacy and security concerns. To address this issue, intense research and development efforts have been directed toward preventing information leaks within the industry. Traditionally, data from the network edges have been sent to centralized clouds for generating accurate and valuable models [3]. However, this approach inherently carries the risk of data leakage during the transmission and storage processes [4].

In response to the privacy and security challenges, regulatory frameworks like the European General Data Protection Regulation (GDDR) now demand service providers to implement technical measures ensuring security, transparency, and accountability. This further emphasizes the criticality of safeguarding personal data and privacy [5], prompting the reevaluation of common practices and potential redesign to address privacy concerns in the context of ML.

To tackle the issue of data exposure during transmission and storage and enhance privacy, Google introduced federated learning in 2016 [6]. Unlike traditional cloud-based ML, federated learning involves conducting a certain amount of training directly at the data source or nearby, thereby reducing the data size sent to the cloud. This approach significantly decreases latency and enhances privacy by ensuring that the data remain at its source [7].

While federated learning addresses privacy concerns, it does not fully resolve the problem of centralization. The weights sent through the network are typically collected at a cloud-based central entity, raising concerns about the stability of cloud service providers and potential model skewing due to favoring certain data streams. Malicious central servers also present risks, and the presence of a centralized curator increases the likelihood of data leakage [8].

To overcome the centralization issues and achieve a distributed consensus, blockchain technologies have been developed. Integrating blockchain in the federated learning process offers a solution to centralized cloud service providers, enabling secure data retrieval and ensuring accurate, traceable, and transparent model training [9].

However, utilizing blockchain for storing and updating the distributed learning model presents its own set of challenges. Data storage on the blockchain can be expensive, limiting the complexity of the federated learning model. Additionally, a robust mechanism is needed to validate incoming data streams to ensure consensus among nodes regarding data validity. Incentivizing participating nodes to perform costly computations during weight training is also crucial for successful implementation [10].

Since Google introduced federated learning, extensive research has been conducted in this area. Previous studies by Bonawitz et al. and Kairouz et al. have explored the overall architecture and design of networks formed by IoT devices connected via the Internet, along with opportunities and threats associated with them [11,12]. Similarly, blockchain technology has seen substantial progress and exploration in various application areas. However, the training of a distributed learning model on a blockchain with a rigorous data verification mechanism has received limited attention to date.

Hence, the main focus of this chapter is to delve into the current state of research on blockchain-based federated learning and provide a proof of concept for such a distributed learning model. This includes the understanding of how blockchains can support federated learning, how models can be trained using the blockchain, and how IoT device data can be integrated into the federated learning model. Addressing common issues with traditional cloud-based ML models, such as data leakage, exposure, centralization, and trust issues, is a key objective of this research. Furthermore, this chapter emphasizes validation and verification of data sent by data sources, with a focus on smart contract (SC)-based verification [13].

In modern distributed networks, federated learning has gained popularity due to its ability to train statistical models directly on remote devices, reducing the need for centralized data storage and ensuring privacy [14]. The process involves constructing a single global statistical model from data stored on numerous remote machines. Device-generated data are stored and processed locally, with only intermediate updates interacting with a central server regularly. At regular intervals, the devices load the current global model and train a local version of the model using their local data. These local models are then updated by the devices based on their respective local data. Subsequently, the local models are sent back to the central server, where they are aggregated into one model, becoming the next iteration of the global model [15].

Overall, this chapter promises to advance privacy-aware ML and explores the potential of blockchain-based federated learning, contributing to cutting-edge technologies and securing sensitive data in the age of IoT and distributed networks.

1.2.1 Federated Learning Algorithm

Google's original federated learning algorithm [14] assumes experimental conditions where edge devices respond, and no malicious devices are present. The algorithm follows a synchronous update mechanism communicated in rounds, involving a collection of K clients, each with its own local dataset. At the beginning of each round, a random percentage C of clients is chosen, and the server delivers the current global algorithm state to them, which represents the current global model [14].

The optimization task in federated learning aims to minimize the overall error, which can be formulated as follows:

$$\text{minimize } f(w) = 1/n \sum n \ f_i(w)$$

where $f_i(w)$ represents the loss function for a specific training sample (x_i, y_i) with the current weights w_i. The loss function measures the error between model predictions and the true input data. The goal is to minimize this error during training [14].

Google's federated learning approach builds upon stochastic gradient descent (SGD), a widely used optimization algorithm in deep learning. The federated optimization uses variants of SGD, where a proportion of clients C is selected in each round, and the gradient of the loss is computed over their local data. C determines the global batch size, and when C = 1, the algorithm is called FederatedSGD (FedSGD) [14].

Another approach, called FederatedAveraging (FedAvg), involves each client performing multiple iterations of computing the local update before averaging over all iterations and obtaining the average of all the client updates [14].

Overall, this research aims to address challenges related to centralization, data privacy, malicious updates, and lazy clients in federated learning to develop a comprehensive and efficient FML architecture for real-world use cases. The proposed solutions will contribute to the broader adoption of FML in various industries and applications, ensuring privacy, security, and accuracy in model training.

1.2.2 Components of a Neural Network

An artificial neural network's neuron shares functional and structural similarities with a biological neuron. It can send its output to multiple other artificial neurons, receiving feature values from external data samples or the outputs of other neurons as inputs. The weighted sum of all inputs is computed for each neuron's output, taking into account the weights of links connecting the inputs to the neuron, along with a bias term. This weighted total, known as the activation, is then passed through an activation function, typically nonlinear, to produce the neuron's final output. The first layer of neurons receives external data as inputs, while the final outputs achieve the network's classification or regression goal [16].

The neural network consists of connections, where each link serves as an input to another neuron. Each link is assigned a weight reflecting the strength of the relationship. A neuron may have multiple input and output connections [17].

Forward propagation is employed in deep learning, where the input to a neuron is computed as a weighted sum of inputs from previous neurons. The inputs are multiplied by the weights of the next layer, and the bias of each layer is added. An activation function is applied if applicable. This process is repeated for each layer until the final output layer is reached [18].

Deep learning networks typically organize neurons into multiple layers, forming complex structures. Neurons in a layer communicate only with neurons in the immediately preceding and succeeding layers. The input layer accepts data from external sources, and the output layer produces the final outcome. Between them, there may be zero or more hidden layers. Networks with a single layer or without hidden layers are also used. Various connection patterns can be established between two layers, such as fully connected neural networks, where every neuron in one layer connects to each neuron in the following layer. Alternatively, pooling may occur, where a group of neurons in one layer connects to a single neuron in the next, reducing the number of neurons in that layer. These connection patterns form a directed acyclic graph known as a feedforward network in mathematics [19].

1.2.3 Learning Process

Learning in a neural network involves improving its ability to perform classification or regression more efficiently based on sample observations. The process revolves around adjusting the network's weights (and optional thresholds) to enhance the accuracy of its output and reduce observable mistakes. Learning is considered complete when additional data do not significantly reduce the error rate, though it never reaches zero. If the error rate remains unacceptably high after learning, the network must be rebuilt. To accomplish this, a cost function is defined and evaluated at regular intervals during the learning process. For classification tasks, the network's output is mapped to specific classes, and a loss function computes the error between predicted and actual labels for input data. The objective of learning is to minimize the sum of errors across all observed data [17].

The learning rate determines the magnitude of corrective steps the model takes for each observation's mistakes. A higher learning rate reduces training time but may lead to lower overall accuracy, while a lower learning rate increases training time but has the potential for higher accuracy. The choice of a cost function is influenced by desired characteristics, such as convexity, or derived from the model, like using a cost function based on the model's posterior probability. The learning rate is essential in preventing the model from getting stuck in local minima and instead converging toward a global minimum [20].

During learning, backpropagation is a technique used to adjust the weights to account for each discovered mistake. It efficiently distributes the total number of errors across connected devices. Backpropagation calculates the gradient (derivative) of the cost function associated with a particular state, enabling weight updates. SGD is a commonly used method for updating the weights [18].

1.2.4 Incentive Mechanism

In order to counteract dishonest behaviors, we have implemented an incentive system with specific objectives. This system aims to motivate workers to act honestly and actively contribute to the task, leading to economic rewards. Our approach involves fostering competition among the workers, prompting each participant to strive for excellence and maximize their potential returns.

1.2.5 Robustness in Federated Learning Systems

Federated learning, a decentralized approach where local learning nodes update a central server without sharing raw data, offers significant privacy benefits. However, it introduces challenges, as deep neural networks are vulnerable to adversarial attacks. To address this, the "ZeKoC" strategy clusters weight updates in a zero-knowledge manner, enhancing robustness against malicious participants.

Another research area focuses on model poisoning attacks, where false data corrupt the model. Tactics like boosting malicious updates and parameter estimation for benign updates are explored to maximize attack effectiveness. "BLADE-FL" model, designed for IoT-based federated learning on a blockchain, mitigates the losses caused by lazy clients through zero-knowledge proofs.

Trusted execution environments (TEEs) provide hardware and software isolation, ensuring privacy for ML applications. Integrating TEEs with deep neural networks, like Intel SGX, showcases their feasibility in privacy preservation. Blockchain technology incentivizes protocol compliance in federated learning, compensating participants with cryptocurrency.

State channels enhance supervision in federated learning, creating sandboxes for trust supervision and data sharing. Various approaches, including ZeKoC, BLADE-FL, TEEs, and blockchain incentives, contribute to privacy, security, and integrity in federated learning systems. Researchers continuously explore novel solutions to tackle the distributed and privacy-sensitive challenges in federated learning.

1.2.6 Current Issues

The field of FML holds tremendous potential, offering several benefits. However, its widespread adoption faces challenges that need to be addressed. This chapter aims to analyze and evaluate these issues to develop a comprehensive FML architecture. The primary challenges include:

1. **Centralization**: In conventional FML, a Central Server or cluster coordinates and aggregates data from clients. The presence of a central aggregator introduces a single point of failure, where any malicious activity or fault can disrupt the entire training process. Deciding on the aggregator's role also presents a challenge.

2. **Data Privacy**: While FML architectures protect client datasets from disclosure, research has demonstrated the possibility of reconstructing some of the original data points from proposed model updates using reverse engineering techniques. Preserving data privacy remains critical.

3. **Malicious Updates**: In a distributed FML environment, malicious clients can intentionally submit incorrect or poisoned updates, potentially sabotaging the global model. This issue becomes especially pronounced in public scenarios where identifying honest and dishonest clients upfront is challenging.

4. **Lazy Clients**: In decentralized FML frameworks, participants can copy the updates submitted by other clients, avoiding the effort of performing any work while still reaping the benefits of honest contributors.

This chapter will delve into these challenges and explore potential solutions, with a focus on creating a robust FML architecture. It aims not only to address private scenarios where corporations or societies collaborate with shared goals but also to cater to public scenarios where anyone interested in contributing to training ML models can participate (e.g., individuals with smartphones contributing to facial recognition software development while preserving their data privacy).

By tackling centralization, ensuring data privacy, preventing malicious updates, and countering lazy client behavior, this chapter endeavors to foster the wide adoption of FML in real-world use cases, unlocking its full potential in various industries and applications.

1.3 Blockchain Technology

Blockchain technology has witnessed significant growth across various fields, showcasing diverse applications, especially since the emergence of Bitcoin. Although the term "blockchain" gained popularity through Ethereum, it originally referred to the foundational infrastructure of the Bitcoin system.

A blockchain is characterized as a decentralized and distributed ledger, where all nodes in a peer-to-peer network possess identical copies of transaction records. These transactions are grouped into blocks, sequentially linked, and undergo validation through consensus mechanisms like Proof of Work (PoW) or Proof of Stake (PoS). In the PoW consensus mechanism, exemplified in Bitcoin, a node initiates a transaction, and other nodes verify it. Miners compete to generate a new block containing validated transactions by finding a unique numerical value (nonce) that meets a specific target requirement. The successful miner receives transaction fees and newly created bitcoins.

Ethereum, with its SC functionality, facilitates the deployment of programmable contracts on its blockchain through the Ethereum Virtual Machine (EVM). Users interact with SCs by submitting transactions executing predetermined functions defined within the contract. The concept of gas holds significance, representing the computational effort required for operations on the Ethereum network. Users pay transaction fees in gas, with the gas price determined in real time based on network congestion.

To address Ethereum's scaling issues and mitigate the impact of rising gas prices, Layer 2 solutions like rollups have emerged. Rollups consolidate multiple transactions into a single transaction on Layer 1, resulting in substantial fee reductions. Optimistic rollups presume transactions are valid unless proven otherwise, while zero-knowledge rollups offer validity proofs without optimistic assumptions. Furthermore, sidechains serve as an alternative scaling solution, operating in parallel with Ethereum, but independent of its security or data availability. Sidechains interact with Ethereum through communication bridges, involving certain trade-offs when compared to Layer 2 solutions.

1.3.1 Ethereum

Ethereum introduced SCs and their deployment on its network, setting a trend for other blockchains to follow suit. This led to Solidity, Ethereum's programming language, which is becoming the most widely used language for designing SCs. The major building blocks of the Ethereum blockchain include states, state transitions, accounts, messages, and transactions [21].

Compared to the Bitcoin blockchain, Ethereum is more computationally capable. While Bitcoin only stores a copy of the transaction list, Ethereum blocks contain both the transaction list and the most recent state, along with additional values such as block number and difficulty. The state transition function is integral to the block validation algorithm, which executes transaction code. Consequently, all nodes executing transaction verification code will also execute contract code as part of the state transition function in the block validation process.

When a transaction is added to a new block (B), the code execution initiated by that transaction is executed by all nodes, both current and future, that download and validate block B [22].

1.4 Privacy Preserving

1.4.1 Zero Knowledge

Zero-knowledge (ZK) proof is a method where a prover convinces a verifier that it possesses specific information (a secret) without revealing the actual information. The zero-knowledge proof must satisfy two crucial requirements: completeness, ensuring an honest verifier is convinced when the statement is true, and soundness, preventing a cheating prover from convincing an honest verifier of a false statement.

For true zero-knowledge, the proof must achieve both completeness and soundness without transmitting the secret information between the prover and verifier. This property makes zero-knowledge proofs useful for privacy-sensitive applications, like authentication systems, where credentials or identities can be verified without disclosing the actual details.

There are two types of zero-knowledge proof approaches: zk-SNARK (Succinct Non-interactive ARgument of Knowledge) and zk-STARK (Succinct Transparent ARgument of Knowledge). zk-SNARKs are well-established and adopted, but they require a trusted setup, potentially leading to centralization issues.

In contrast, zk-STARKs do not require a trusted setup, but their proofs are larger compared to zk-SNARKs. The main advantage of zk-STARKs is their transparency, while zk-SNARKs have computational soundness.

1.4.1.1 Interactive Proofs Zero-knowledge proofs require interactive engagement between entities proving their knowledge and validators. The prover responds to challenges posed by the verifier, convincing them of the validity of a statement only if the prover possesses the claimed knowledge. To avoid protocol replay attacks, the verifier should not learn any additional information beyond the truth of the statement.

Three essential properties of zero-knowledge proofs must be met:

1. **Completeness**: An honest prover convinces an honest verifier if the statement is true.
2. **Soundness**: No cheating prover can persuade an honest verifier that a false statement is true, except with an extremely small probability.
3. **Zero Knowledge**: The verifier gains no extra information about the secret, apart from the statement's truth.

The third property distinguishes zero-knowledge proofs from regular mathematical proofs, making them probabilistic. Although there is a small chance of a soundness error, strategies can be employed to minimize this possibility to negligible levels.

1.4.1.2 Non-Interactive Proofs In 1988, Blum, Feldman, and Micali demonstrated the possibility of achieving computational zero-knowledge without interaction by using a common reference string shared between the prover and the verifier. This breakthrough eliminated the need for back-and-forth communication. However, in the conventional model, Goldreich and Oren later proved that one-shot zero-knowledge procedures are not possible.

The characteristics exhibited by a zero-knowledge protocol are influenced by the model used. Non-interactive zero-knowledge protocols, as studied by Pass, do not preserve all the qualities of interactive zero-knowledge protocols, such as deniability, under the common reference string model.

In 2012, Bitansky et al. introduced the term "zk-SNARK," which stands for zero-knowledge succinct non-interactive argument of knowledge. This concept gained prominence in the Zerocash blockchain system, where zero-knowledge cryptography played a crucial role in enabling mathematical proofs of possession without revealing the actual information. The zk-SNARKs found widespread deployment in this context.

Non-interactive zero-knowledge proofs are particularly useful when a large number of observers need to efficiently verify the proof. However, it is worth noting that zero-knowledge proofs don't necessarily have to be non-interactive all the time. In some cases, it might be more practical to identify a reliable validator who can vouch for the integrity of the proof.

The Fiat–Shamir heuristic is a technique used to derive a digital signature from an interactive demonstration of knowledge. This allows validation of a public fact or witness without requiring the prover to be continuously available for interaction.

1.4.2 Differential Privacy

Differential Privacy (DP) serves as a method to share information about a dataset publicly while safeguarding the privacy of individual data points. It achieves this by revealing general statistics and behavioral patterns of the dataset's population, without disclosing specific information about any single individual. The fundamental idea behind DP is to introduce random noise or substitutions to the data in such a way that the overall impact on the population's average behavior remains minimal, thereby ensuring privacy.

An algorithm is considered differentially private if an observer examining its output cannot determine whether a particular individual's data were used during the computation. In the context of this chapter, DP is employed as a means to prevent the reconstruction of original data points used in updating the model. This is achieved by injecting random noise into the datasets before the training process. When the noise level is appropriately set, it becomes highly challenging for an attacker to determine whether a specific data point from the target victim was used or not.

However, it's essential to be cautious while tuning the added noise, as it can impact the precision of the model update obtained.

1.5 Blockchain-Enabled Federated Learning

The IoT has witnessed continuous growth, resulting in vast amounts of data that greatly contribute to the success of current AI systems. To optimize the utilization of these data, prevailing distributed learning systems traditionally train their models on a centralized cloud infrastructure with substantial computing capacity.

Amidst the rising popularity of IoT networks, concerns regarding privacy and centralization have become prominent. Data generated by IoT devices often contain highly sensitive information, making it vulnerable during transmission or storage on cloud servers. In response to this, federated learning was introduced, preserving sensitive data on devices to enhance privacy.

However, the concentration of cloud computing services limits options for IoT devices in federated learning, impacting the selection of suitable providers with necessary aggregation capabilities. Moreover, potential risks arise from cloud service providers engaging in fraudulent or malicious behavior, compounded by the lack of visibility into cloud-based training techniques. As a consequence, data providers remain unaware of the utilization of their data once the model is trained.

To address the challenge of non-transparency, this chapter employs a blockchain-based federated learning paradigm. SCs deployed on the blockchain, with code linked to the contract's address, ensure transparency for all training participants. Despite the advantages of blockchain-based federated learning, certain limitations persist in establishing the legitimacy and involvement of training participants. Dishonest behavior to conserve computing resources is a concern. To counter this, this chapter proposes using zero-knowledge proofs to impose significant penalties for malicious behavior, encouraging greater honesty among participants.

1.5.1 Current Solutions

In recent efforts to improve federated learning through blockchain integration, the focus is on enhancing decentralization and mitigating potential malicious behavior. Blockchain plays a key role in secure model parameter storage, auditing, and transparent exchange among participants. Zero-knowledge proofs are utilized to handle malicious activities seamlessly and securely.

An alternative approach called on-device federated learning accumulates local model updates on the blockchain, aiming to address centralization and provide motivational incentives for participants. However, this chapter emphasizes a more purely decentralized approach with minimal compromises to ensure scalability.

The main limitation of FML lies in centralization, prompting the emergence of decentralized solutions incorporating blockchain, a distributed ledger technology. Blockchain facilitates trustless and decentralized coordination through SCs, code executed automatically when specific conditions are met. While custom blockchain solutions offer communication efficiency and tailored solutions, they lack extensive testing compared to existing public blockchains, making the latter a preferable choice for achieving decentralization in FML.

Frameworks using existing blockchains like Ethereum face challenges due to the computation and storage limits of SC-executed code. To overcome this, some approaches implement chunking to divide the code into multiple SCs, yet this method leads to complexity and high transaction costs. To address these issues, the proposal suggests moving the FedAvg algorithm's averaging part to the clients and using a decentralized storage solution for their model updates.

Data privacy is addressed using DP, which adds random noise to datasets during training to safeguard against reverse engineering of updates. Although this slightly reduces model accuracy, it protects data privacy. To handle malicious updates, some solutions propose a common test set for assessing update quality, but this approach results in significant performance overhead and risks misuse. An incentive system is considered more effective in discouraging malicious behavior.

The proposed solution builds on the work of Toyoda et al. [23], dividing clients equally among rounds and employing blockchain instead of central servers. Clients evaluate and vote for the best updates from the previous round, average the selected models, and conduct local training. The algorithm, referred to as k-Uniform FedAvg, rewards clients submitting the best-voted models through an incentive system based on Mechanism Design. The rewards are funded by entities interested in the task's success.

1.5.2 Different Use Cases

1.5.2.1 An Assisted Diagnosis Model for Cancer Patients Based on Federated Learning This research focuses on the development of a federated learning system aimed at diagnosing cancer recurrence and determining the tumor location after surgery. The primary concern lies in the likelihood of tumor recurrence after surgery, making the prediction of recurrence time critical for effective patient treatment. Additionally, distinguishing whether the tumor has reappeared at the same site or has metastasized holds significant importance.

To achieve this, the researchers employed Convolutional Neural Networks (CNN) as local learning models. They updated the global model by transmitting the weights and biases of these local learners to a central server. To safeguard the privacy of patients' data, a localized DP mechanism was established on the user side.

Utilizing the federated learning model, they effectively addressed the issue of data scarcity, leading to improved accuracy. Figure 1.1 illustrates the structure of their federated learning architecture.

Additionally, Figure 1.2 provides a clear overview of their proposed structure.

Furthermore, the pseudo-code for their proposed system is depicted in Figure 1.3.

1.5.2.2 Swarm Learning for Decentralized and Confidential Clinical ML This research introduces a novel framework for distributed learning called "Swarm learning." Unlike federated learning, where there is a central aggregation point, Swarm learning omits the central aggregation and facilitates communication between nodes for sharing their local models. In this approach, instead of uploading weights and biases to an SC, the nodes directly communicate with each other and update their local weights.

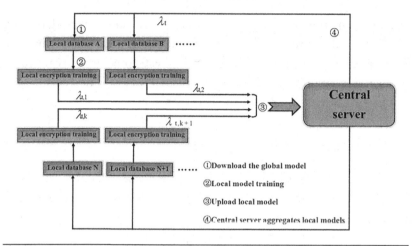

Figure 1.1 Federated learning architecture.

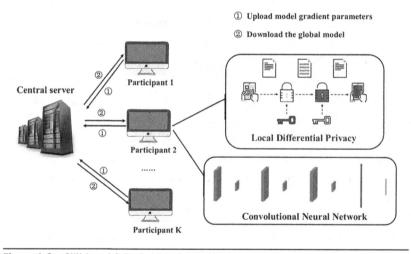

Figure 1.2 CNN-based federated learning architecture.

Figure 1.4 illustrates the structure of this Swarm learning framework along with other types of learning frameworks.

1.5.2.3 Federated Learning Applications in Health We also explore three distinct studies that leverage federated learning (FL) in different health-related domains while incorporating blockchain technology for enhanced security and privacy.

Algorithm 1 CNN-FL model based on local differential privacy

---

1 For Iteration t do
/* Service-Terminal: */

2 $\omega_t = \dfrac{1}{Q}\sum\limits_{q=1}^{Q}\omega_t^k;$

3 Send ω_t to each participant;
/* participants: */
4 for Participant q do
5 Do Localized differential privacy
6 $\omega_t^q = \omega_t;$
7 For Local epoch e do

8 $\omega_t^q = \omega_t^q - \eta\dfrac{\partial}{\partial X^q}Z_{\wp}$

9 End
10 End
11 End

Figure 1.3 CNN-FL model pseudo-code.

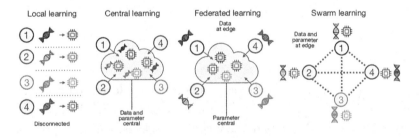

Figure 1.4 Swarm learning framework architecture.

The first study focuses on genome-wide association studies, introducing a unique combination of genomic data and blockchain technology. Blockchain is utilized to prevent unauthorized access to sensitive genomic information, using an SC to maintain a list of authorized users for efficient identity verification. Federated learning is employed to collaboratively train ML models without sharing raw data, ensuring privacy preservation.

The second study explores the applications of federated learning in predicting cancer using CT images. Privacy concerns are addressed through the federated learning approach, enabling multiple health-care institutions to collaborate and train predictive models without sharing sensitive patient data. While genomic data are not the focus in these studies, it is highlighted that deep learning techniques have been commonly employed as the learning method in various cancer diagnosis applications using federated learning.

The third study employs a federated learning approach for diagnosing Parkinson's disease. Training data comprise tomography images, fed to a CNN network. The global model is stored on a central server, accessed by clients through a blockchain-based mechanism. Unlike the first study, the global model distribution is managed by the central server instead of an SC. Clients securely send gradient descent information instead of weights and bias matrices to update the global model, preserving privacy.

Overall, these studies demonstrate the promising potential of federated learning in diverse health applications, enabling collaborative model training while ensuring data privacy and security through blockchain integration. The architectures of the proposed systems are noteworthy, providing valuable insights into the use of federated learning and blockchain technologies in the healthcare domain.

1.6 Conclusion

This chapter explores FML as a solution to address data privacy concerns in ML applications. The main obstacles identified include centralization, malicious and lazy clients' updates, and data privacy.

To tackle the issue of centralization, blockchain architectures were explored for federated learning, proving to be effective. The approach built on the SC concept made it compatible with Ethereum and other EVM-compatible blockchains, overcoming computation and storage limitations. The role of coordinator is assumed by an SC, and decentralized storage is used to store model updates, while participants in the FML task perform update aggregation.

A key contribution is the ability to prove whether a client effectively performed the required training procedure of a neural network in a federated learning solution, achieved through zero-knowledge STARK proofs.

The evaluation of the blockchain-based federated learning model with zero-knowledge proof data verification demonstrates performance comparable to the theoretical limit. While the predictability of different classes varies, given sufficient time and data, all labels show a convergence pattern.

However, integrating blockchain and zero-knowledge proof introduces significant overhead, making the process slower, and requiring consideration of gas costs when determining the interval time for training rounds. Increased data volume per update round results in more time needed to generate proofs, impacting the verification process. Nonetheless, local training time with more data points remains negligible.

The system exhibits decentralization, performance, verifiability, robustness, and privacy. However, achieving these traits often comes with either time or monetary costs.

References

[1] Q. Yang, Y. Liu, T. Chen, and Y. Tong. "Federated Machine Learning: Concept and Applications". ACM Transactions on Intelligent Systems and Technology 10.2 (2019), pp. 1–19.

[2] Stefano Savazzi, Monica Nicoli, and Vittorio Rampa. "Learning with Cooperating Devices: A Consensus Approach for Massive IoT Networks". IEEE Internet of Things Journal 7.5 (2020), pp. 4641–4654. doi: 10.1109/jiot.2020.2964162.

[3] Erwin Adi et al. "Machine Learning and Data Analytics for the IoT". May 2020. doi: 10.1007/s00521-020-04874-y

[4] Wei Wu et al. "Blockchain Based Zero-Knowledge Proof of Location in IoT". In: 2020 IEEE International Conference on Communications (ICC) (2020). doi: 10.1109/icc40277.2020.9149366.

[5] "Data Protection in the EU". Jan. 2021. https://ec.europa.eu/info/law/law-topic/data-protection/data-protection-eu_en

[6] Felix Sattler et al. "Robust and Communication-Efficient Federated Learning from Non-IID Data". In: CoRR abs/1903.02891 (2019). arXiv: 1903. 02891. http://arxiv.org/abs/1903.02891.

[7] Yuzheng Li et al. "A Blockchain-Based Decentralized Learning Framework with Committee Consensus". IEEE Network 35.1 (2020), pp. 234–241. doi: 10.1109/mnet.011.2000263.

[8] Jun Li et al. "Blockchain Assisted Decentralized Learning (BLADE-FL) with Lazy Clients". 2020. arXiv: 2012.02044 [cs.LG].

[9] Yunlong Lu et al. "Blockchain and Learning for Privacy-Preserved Data Sharing in Industrial IoT". IEEE Transactions on Industrial Informatics 16.6 (2020), pp. 4177–4186. doi: 10.1109/tii.2019.2942190.

[10] Umer Majeed and Choong Seon Hong. "FLchain: Learning via MEC-enabled Blockchain Network". In: 2019 20th Asia-Pacific Network Operations and Management Symposium (APNOMS) (2019). doi: 10.23919/apnoms.2019.8892848.

[11] Keith Bonawitz et al. "Towards Federated Learning at Scale: System Design". In: CoRR abs/1902.01046 (2019). arXiv: 1902.01046. http://arxiv.org/abs/1902.01046.

[12] Peter Kairouz et al. "Advances and Open Problems in Federated Learning". In: CoRR abs/1912.04977 (2019). arXiv: 1912.04977. http://arxiv.org/abs/1912.04977.

[13] H. Pranith et al. "End-to-End Verifiable Electronic Voting System Using Delegated Proof of Stake On Blockchain". SSRN Electronic Journal (2019). doi:10.2139/ssrn.3511409.

[14] H. Brendan McMahan et al. "Federated Learning of Deep Networks using Model Averaging". In: CoRR abs/1602.0562.

[15] Jakub Konečn˝ et al. "Federated Learning: Strategies for Improving Communication Efficiency". In: CoRR abs/1610.05492 (2016). arXiv: 1610.05492. http://arxiv.org/abs/1610.05492.

[16] Maysam F. Abbod et al. "Application of Artificial Intelligence to the Management of Urological Cancer". Journal of Urology 178.4 (2007), pp. 1150–1156. doi:10.1016/j.juro.2007.05.122.

[17] Andreas Zell et al. "SNNS (Stuttgart Neural Network Simulator)". In: Neural Network Simulation Environments. Ed. by Josef Skrzypek. Boston, MA: Springer (1994), pp. 165–186. doi:10.1007/978-1-4615-2736-7_9.

[18] Christian W. Dawson and Robert Wilby. "An Artificial Neural Network Approach to Rainfall-runoff Modelling". Hydrological Sciences Journal 43.1 (1998), pp. 47–66. doi:10.1080/02626669809492102.

[19] M. Fredrikson, S. Jha, and T. Ristenpart. "Model Inversion Attacks That Exploit Confidence Information and Basic Countermeasures". In: Proceedings of CCS. New York, NY: ACM (2015), pp. 1322–1333.

[20] Yong Li et al. "The Improved Training Algorithm of Back Propagation Neural Network with Self-adaptive Learning Rate". In: 2009 International Conference on Computational Intelligence and Natural Computing. Vol. 1. (2009), pp. 73–76. doi:10.1109/CINC.2009.111

[21] Gavin Wood et al. "Ethereum: A Secure Decentralised Generalised Transaction Ledger". In: Ethereum Project Yellow Paper 151.2014 (2014), pp. 1–32.

[22] Ethereum Whitepaper. https://ethereum.org/en/whitepaper/.

[23] K. Toyoda, J. Zhao, A. N. S. Zhang, and P. T. Mathiopoulos. "Blockchain Enabled Federated Learning with Mechanism Design". IEEE ACCESS (2020), pp. 219744–219756. doi:10.1109/ACCESS.2020.3043037.

2

EMPOWERING HEALTHCARE

Symbiotic Innovations of AI and Blockchain Technology

MARUF FARHAN

Figure 2.1 Future frontiers: Digital pioneers of AI and blockchain in healthcare.

DOI: 10.1201/9781003162018-2

2.1 Introduction

There's no denying that artificial intelligence (AI) has risen to prominence in recent years and that there's a great deal of interest in the field among academics and businesses. John McCarthy, an MIT computer science professor, created AI. He began his groundbreaking work on AI by studying how computers could mimic human actions. McCarthy is often cited as one of the pioneers in AI because he popularized the phrase "artificial intelligence" in 1955. He theorized that machines might replace human labor and provided examples in his writings. McCarthy thought machines could acquire knowledge by observation and progress as time passed. He predicted that technology would one day replace humans in law enforcement and medicine Figure 2.1. Since then, researchers have expanded the use of AI technologies across numerous industries, from healthcare to robots to banking, speech recognition, and beyond. Early in the 1960s, it became obvious that machines could learn and increase their abilities over time. AI is currently being hailed as one of the most revolutionary technologies in recent times, with the potential to alter how we live and work completely.

AI refers to the capability of a machine or a computer program to perform tasks that would normally require human intelligence, such as visual perception, speech recognition, decision-making, and natural language processing (NLP) [1].

As many sectors implement AI-based solutions to boost productivity and precision, Gartner predicts that the worldwide business value of AI will reach $3.9 trillion by 2022 (Table 2.1).

Now, let's start with a definition of blockchain technology. Did anyone ever realize that we can make financial transactions without support from a bank or any other intermediaries? In the era of technology, this million-dollar question was answered in October 2008. It was a revolutionary invention of this millennium, no doubt.

Table 2.1 AI-Based Solutions Business Value [2]

	2017	2018	2019	2020	2021	2022
Business Value	692	1,175	1,901	2,649	3,346	3,923
Growth (%)		70	62	39	26	17

With the term "Blockchain," now we can imagine a world where contracts are written in some codes and kept secret in a database that gets protection from all sorts of tempering, deletion, or revision.

2.1.1 Let's Help You Understand the Basic Concept of Blockchain with a Simple Example

Consider a group of mates interested in keeping track of their joint spending. They could keep track of everything with a spreadsheet or other online spreadsheet document, but then someone would have to administer it and ensure the data are valid. They opt to employ a blockchain instead. A consensus protocol for adding new transactions to the blockchain was discussed, and it was stated that "each friend is going to have their copy of the blockchain."

In this example, we'll pretend that Alice owes Bob $10. Alice inserts this transaction into a new block, and her friends confirm that it is legitimate (i.e., that she does owe Bob $10). After the transaction has been validated, a new hash is generated for the block, which is added to each friend's copy of the chain.

Let's pretend that Bob has $5 and is interested in purchasing something from Charlie. Bob includes this transaction in a new block, and his pals check to ensure it's legitimate before accepting it. An identical copy of the chain is constructed for each friend, and a new hash is generated for the newest block.

The more transactions there are in a chain, the more secure and difficult to alter each block will become. This is because each block includes a distinct hash calculated from the block's contents and the preceding block's hash. All subsequent hashes would become invalid if a block were tampered with, sending a signal to the network that something was amiss.

The above scenario demonstrates the basic idea of a blockchain. With the help of cryptography and a decentralized network (a network with no central authority or control), blockchain can provide a secure and transparent way for people to make financial transactions.

Now let's get back to the history of blockchain again. In 2008, a whitepaper called "Bitcoin: A Peer-to-Peer Electronic Cash System" by an anonymous person or group of people using the name "Satoshi Nakamoto" was the first time the idea of blockchain was mentioned [3].

Table 2.2 Summary of Blockchain Technology

YEAR	DEVELOPMENT
2008	Satoshi Nakamoto published the Bitcoin whitepaper describing the concept of a decentralized digital currency built on a blockchain.
2009	The Bitcoin blockchain is launched, and the first block, the "genesis block," is mined.
2011	Litecoin and Namecoin are new blockchains that slightly modify the Bitcoin blockchain system.
2014	The Ethereum blockchain is introduced, offering a more versatile and programmable blockchain that allows for the creation of smart contracts and decentralized applications (DApps).
2015	The Hyperledger project is a joint effort by the Linux Foundation and several industry partners to develop an open-source and enterprise-ready blockchain infrastructure.
2016	To investigate the potential uses of blockchain technology in the financial sector, some 80 financial institutions banded together to form the R3 Consortium, the first public blockchain collaboration.
2017	To raise capital, a growing number of companies are turning to initial coin offerings (ICOs) in the form of the sale of cryptocurrency tokens.
2018	The InterWork Alliance is a nonprofit organization creating universal blockchain protocols and specifications.
2020	The Ethereum 2.0 upgrade has begun, and its purpose is to increase the scalability, security, and durability of the Ethereum network.

Even though no one knows who Satoshi Nakamoto is, their work on the Bitcoin blockchain significantly impacted how blockchain technology developed.

Table 2.2 shows a quick summary of how blockchain technology has changed over time.

2.1.2 Blockchain Layered Technology

2.1.2.1 Layers of Blockchain Technology A blockchain system has several interconnected layers to securely store information, reach consensus, and execute smart contracts. Blockchain layers often include the following:

1. **Application Layer:** The application layer is a blockchain's last and most complex part. It's made up of apps that people use to communicate with the distributed ledger. Examples include wallets, user interfaces, and decentralized apps (DApps). Users can access the blockchain network and its features via the application layer [2].

2. **Smart Contact Layer:** This layer enforces and executes the blockchain network's business logic and governing laws. "Smart contracts," or contracts with pre-programmed triggers and responses, ensure their conditions are always carried out. They make building decentralized apps on the blockchain possible, facilitating tasks' automation and simplifying financial transactions [1].

3. **Consensus Layer:** This layer coordinates the consensus among blockchain nodes on the integrity and chronological order of transactions. Besides that, it guarantees that the network agrees on the blockchain's current state. Proof-of-Work (PoW), Proof-of-Stake (PoS), and Practical Byzantine Fault Tolerance (PBFT) are just a few examples of consensus processes used to keep the network safe and secure [3].

4. **Network Layer:** This layer handles the communication and connection to the blockchain network. It also ensures data transmission, messages, and transactions between nodes in a peer-to-peer manner [4]. The network layer establishes and maintains the network topology, manages node discovery, and handles data propagation across the blockchain network.

5. **Data Layer:** This layer is used for storing and managing the data of the blockchain. In the distributed ledger, where the entire transaction history is saved, this layer consists of that ledger. It uses cryptographic techniques to integrate the data [5].

2.1.3 The Adaption of AI and Blockchain Technology

The adoption of blockchain and AI in healthcare has the potential to revolutionize the industry and address many of its long-standing issues. However, as with any emerging technology, several challenges must be addressed to ensure its successful implementation. Some of these challenges include patient privacy, providing accurate medical records, and organizing the distribution of medications. Patient privacy is a critical concern in healthcare. Medical records contain sensitive information about patients, including their health conditions, treatments, and personal information. Blockchain technology can address this issue by providing a secure and decentralized platform for storing medical records. Blockchain's distributed ledger technology ensures patient data is encrypted and tamper-proof, reducing the risk of unauthorized access or data breaches. Blockchain also gives patients greater control over their data, enabling them to grant or revoke access to their medical records as they see fit.

Ensuring accurate medical records is another challenge that can be addressed with blockchain and AI. Medical records are often fragmented and dispersed across healthcare providers, making accessing and sharing information difficult. Blockchain's distributed ledger technology enables healthcare providers to access a patient's medical history in real time, regardless of where the records were initially created. AI can also analyze medical records and identify inconsistencies, ensuring patients receive the most accurate diagnoses and treatments.

Organizing the distribution of medications is another problem that blockchain and AI can address. The pharmaceutical supply chain involves multiple parties, including drug manufacturers, distributors, and pharmacies. Blockchain's distributed ledger technology can provide greater transparency and traceability in the supply chain, reducing the risk of counterfeit drugs and ensuring that medications are delivered to patients promptly and efficiently [6]. AI can also optimize medication distribution, predict demand and supply patterns, and improve inventory management [6].

Despite the potential benefits of blockchain and AI in healthcare, several challenges must be addressed before widespread adoption can occur. These challenges include regulatory issues, technical limitations, data ownership and interoperability concerns [7]. However, with continued research and development, blockchain and AI will likely play a significant role in improving healthcare outcomes in the years to come.

2.1.4 Overview of the Benefits of AI and Blockchain Technology in the Healthcare Industry

AI and blockchain technology have attracted great interest in recent years as potential ways for the healthcare industry to enhance efficiency, cut costs, and better serve patients. Blockchain technology can provide a secure and transparent platform for sharing patient data and protecting privacy. Decreasing fraud, while AI can enable healthcare providers to make more accurate diagnoses, design individualized treatment plans, and automate administrative work. This chapter will introduce AI and blockchain technology and explain how these innovations revolutionize healthcare delivery.

There is potential for blockchain technology to be used in medical research. By archiving consumers' informed consent, medical researchers can provide a more open, traceable, and tamper-proof study process [8]. The use of blockchain technology has increased and sectors like financial services [9], the supply chain industries [10], payment gateways industries [11,12] and e-commerce [13] are the ones that have used blockchain technology and received benefits.

2.2 Benefits of AI in the Healthcare Industry

1. **Enhanced Diagnostic:** X-rays, CT scans, and MRIs are only some of the medical images that may be analyzed by AI algorithms with great precision [14].
2. **Personalized Treatment:** To create individualized treatment plans and forecast patient outcomes, AI can analyze enormous volumes of patient data, including medical records, genetic information, and lifestyle data. This may result in more precise and efficient treatments [15].
3. **Discovery of Drugs and Development:** New medicines can be developed more quickly thanks to AI's capacity to analyze large molecular databases, anticipate drug interactions, and accurately identify viable drug candidates [16].

Financially, combining AI and Blockchain technology creates a tremendous financial impact on the healthcare industry. The global market for blockchain technology in healthcare is expected to develop at a compound annual growth rate (CAGR) of 71.9% from 2018 to 2023, with a market value of $5.61 billion by 2023, according to a report published by Market Research Future [17]. According to a second analysis by McKinsey & Company [18], AI can save the US healthcare system $100 billion annually by 2026. However, it is essential to note that the financial impact might vary substantially depending on the application and utilization of AI and blockchain technology in the healthcare industry.

2.3 AI and Blockchain Technology in Healthcare

AI and Blockchain are emerging technologies that can transform the healthcare industry. AI can improve diagnosis and treatment, while blockchain can help secure medical records and ensure data privacy.

One application of AI in healthcare is using machine learning algorithms to analyze large amounts of medical data and identify patterns that could be used to improve patient outcomes. For example, AI could identify patients at risk of developing a particular condition or suggest treatment options based on a patient's medical history. Blockchain technology, on the other hand, can be used to secure and manage medical records. This is particularly important in the context of electronic health records (EHRs), which are becoming increasingly common. By using blockchain to store and manage EHRs, patients can have greater control over their data and ensure that their records are accurate and current. A standardized EHR system is an example of a centrally located design that allows for centralized monitoring, coordination, and direction of the network. AI can analyze vast troves of patient records and perform intricate computations. Some medical professionals hesitate to use AI to improve patient care, even though the technology can already outpace humans in various dynamic and cognitive activities [19,20].

Both AI and blockchain have demonstrated their development in the healthcare sector. As a result, the scientific community and researchers are interested in this field by using this technology. However, blockchain has seen better adoption in the healthcare sector. It improves the security risks associated with the incompatibility of existing EHR systems. Information about patients is managed, collected, organized, and transmitted through a health information system. All the data from the hospital archive are stored in the administrative system. After processing, the information system sends the results to the patient management, clinical information systems, and clinical support services. In addition, it feeds information into an electronic record that physicians can view at the bedside. However, clinical support system tools indicate the next steps in therapy and bring to the patient's attention data the patient may have missed otherwise. Figure 2.2 depicts the EHR architecture.

Another potential application of blockchain in healthcare is in clinical trials. Blockchain can be used to create a secure and

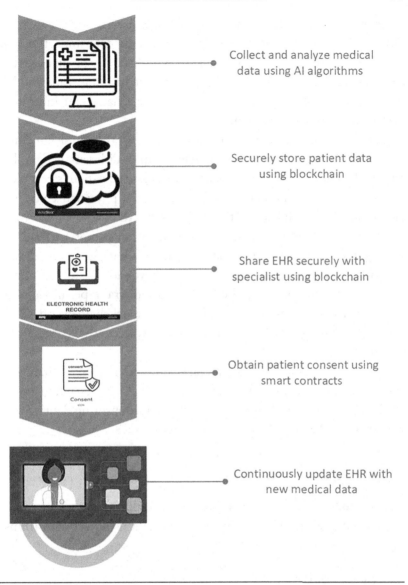

Collect and analyze medical data using AI algorithms

Securely store patient data using blockchain

Share EHR securely with specialist using blockchain

Obtain patient consent using smart contracts

Continuously update EHR with new medical data

Figure 2.2 Proposed framework to improve the health sector by using AI and blockchain.

transparent system for managing clinical trial data, which can help ensure that trial results are accurate and reliable.

Overall, the combination of AI and blockchain has the potential to revolutionize the healthcare industry by improving patient outcomes, enhancing data security, and enabling more efficient and effective healthcare delivery.

Let's say a patient is diagnosed with a rare disease, and the treating physician needs to consult with a specialist to determine the best course of treatment. The physician could use AI to analyze the patient's medical records, including diagnostic imaging scans and lab results, to identify any patterns or anomalies that could indicate a particular course of treatment.

Once the AI has made its recommendations, the physician could consult with the specialist, who may be in a different geographic region. To ensure that the specialist can access the patient's complete medical history, including the AI analysis, the physician could use blockchain to securely share the patient's EHR.

Blockchain would provide a secure and transparent method for sharing the EHR, ensuring that the patient's data is protected and that the physician and the specialist have access to the same information. Additionally, the patient could grant permission for the specialist to access their EHR using a blockchain-based smart contract, ensuring that the specialist can only access the data that the patient has explicitly authorized.

Overall, this example demonstrates how AI and blockchain can work together to improve patient outcomes by enabling more efficient and effective healthcare delivery, while ensuring data privacy and security (Table 2.3).

2.3.1 A Proposed Framework for the Above Example Could Be as Follows

Collect and analyze medical data: The healthcare provider would collect and analyze the patient's medical data using AI algorithms,

Table 2.3 Benefits of AI and Blockchain in Healthcare

AI APPLICATIONS IN HEALTHCARE	BLOCKCHAIN APPLICATIONS IN HEALTHCARE
Improving diagnosis and treatment	Securing and managing medical records
Identifying patients at risk of developing a condition	Creating a secure and transparent system for managing clinical trial data
Suggesting treatment options based on a patient's medical history	Enabling patients to have greater control over their data
Analyzing large amounts of medical data to identify patterns	Ensuring the accuracy and reliability of clinical trial results

which would help to identify any patterns or anomalies that could be used to determine the best course of treatment.

Securely store patient data using blockchain: The healthcare provider would securely hold the patient's EHR on a blockchain network, ensuring that the data are protected and can only be accessed by authorized parties.

Share EHR securely with a specialist: The healthcare provider would use blockchain to securely share the patient's EHR with the specialist, who could then review the AI analysis and other relevant medical data to help determine the best course of treatment.

Obtain patient consent using smart contracts: The healthcare provider would obtain the patient's permission to share their EHR with the specialist using a blockchain-based smart contract, ensuring that authorized parties only access the patient's data.

Continuously update EHR: The healthcare provider would constantly update the patient's EHR with new medical data as it becomes available, which would help to ensure that the patient's medical history is complete and up to date.

The proposed framework advises the audience to use blockchain and AI to build a decentralized, secure, data-driven healthcare system that can enhance healthcare outcomes, lower healthcare costs, and broaden healthcare access.

2.4 An Analysis of the Present State of Healthcare AI and Blockchain-Based Solution Development

2.4.1 Present State of Application—the Integration of AI and Blockchain in Healthcare

The healthcare industry is progressively using AI and blockchain-based technologies to enhance patient outcomes, simplify administrative tasks, and safeguard private information. Recent years have seen a dramatic increase in research and development activities into the possible applications of these technologies in healthcare. Massive volumes of healthcare data, such as medical imaging, EHRs, and genetic information, are being analyzed using AI methods like machine learning and deep learning to glean useful insights.

2.4.2 Digital Health Records and Medical Records

EHRs are digital versions of a patient's medical record containing detailed information about their health and medical history. The system is designed to store, manage, and share this information among the health and medical workers involved with patient care. It needs to be kept secure because they have some vital information about the patients, known as personally identifiable information (PII). If this is exposed, it can lead to security breaches and data privacy breaches [21]. However, many EHR systems struggle with safety, privacy, and management [22]. To fix this problem, blockchain technology can be used to manage and maintain EHR. Researchers proposed a framework where they could store medical history in "smart contracts" [23]. The administrator allocated a position to the user, who may be a doctor or a patient, using Ethereum as the blockchain network. Patient data was added, viewed, and deleted directly to the blockchain network, ensuring security. Also, researchers proposed **bheem**, which is a blockchain-based system to manage HER [24] where Nodes representing patients, healthcare providers, and proxies were all part of the system. This initiative aimed to simplify and protect data access for patients, healthcare professionals, and interested parties.

When it's associated with AI, the HER becomes very useful. Data from EHRs serve a vital role in enhancing the performance and accuracy of acute critical illness prediction, building upon traditional Early Warning Systems (EWS) [25].

As shown in Figure 2.3, a provider node adds a new patient's EHR to the blockchain, and the patient confirms receipt of the record.

AI can review doctors' notes and provide insights. Algorithms can use EHR to predict the likelihood of a disease based on historical data and family history. To train AI systems, scientists employ massive amounts of data, from which the algorithm draws rules that tie its observations to the final diagnosis.

EHRs can be easily licensed to other medical institutions or third-party platforms by its users. Since the doctor only has read access, the client must give permission to upload new permissions and amend existing ones. Physician leadership endorses this patient-centric approach.

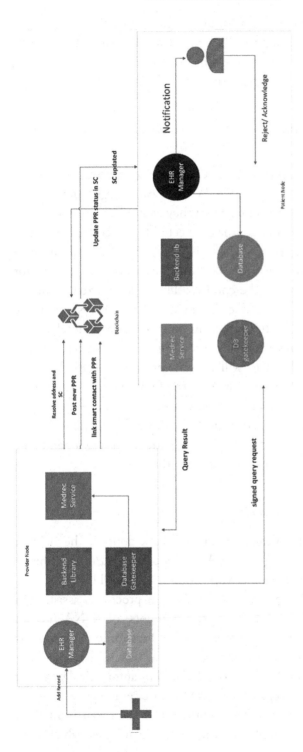

Figure 2.3 Provider adds an EHR for new patients using blockchain.

Interoperability in the identification and verification of patients and medical staff is a real possibility thanks to blockchain technology. For the sake of patient security and medical excellence, a blockchain-based network routinely verifies doctors' credentials [26].

2.4.3 Telehealth, mHealth, Remote Patient Monitoring, and Individualized Medical Care

Following the global spread of the COVID-19 epidemic, telehealth and remote patient monitoring (RPMO) became urgent necessities. Most of them are now centralized and they lack privacy, security, and directness [27]. When a patient is connected to a medical team via a public cloud service, security risks rise. This is especially the case when a patient from a distance location must travel to see a doctor located in another part of the country. The decentralized nature of blockchain technology may one day make telehealth systems more efficient by allowing for the provision of healthcare remotely. This will provide the system with the required transparency while reducing the likelihood that the data will be tampered with [28]. When a patient needs to be isolated, as in the case of a COVID-19 infection, mobile health activities can be carried out in a virtual setting [29]. A blockchain-based diabetic consortium has been formed, and its use could make it easier to prioritize diabetic patients during a pandemic and provide healthcare on a distant basis [30].

2.4.4 Bioinformatics and Genomics

There has been a significant amount of work done in the field of bioinformatics in the categories of AI and machine learning. A high-order convolutional neural network, abbreviated as HOCNN, was developed with the intention of facilitating the process of assisting in the prediction of the sequence requirements of DNA-protein binding [31]. This network implemented high-order associations between nucleotides by making use of a high-order encoding approach to accomplish this goal [32]. To automatically assign ICD-9 codes, the DeepLabeler architecture employs a method which combines CNN with the Document to Vector transformation. To manage the most complex task in sequence analysis, which is to discover protein distant homology, a predictor that is based on Long-Short Term Memory (LSTM) was proposed in [33].

2.4.5 Drug Discovery and Management

Drug development and production are two other areas where deep learning has proven useful. When compounds have uniform cell definitions, fully integrated DNNs make model development easy [30]. The research conducted by DNN in the Merck Kaggle challenge database makes extensive use of 2D topological descriptions, and DNN has done marginally better than RF on 13 out of 15 targets. According to the findings of research, their DNN models for a variety of jobs won Tox21. AI can assist in speeding up the discovery of new medicines by identifying promising new drug candidates and facilitating their rapid evaluation for safety and efficacy. Applying AI has led to new treatments for diseases including Alzheimer's and cancer [34].

2.4.6 Clinical Trials

Clinical trials can become both more transparent and safer when blockchain technology is utilized. TrialX is a firm that is utilizing blockchain technology to provide a decentralized platform for clinical trials [35]. On this platform, patients have the option of maintaining ownership of their data, and the results of the clinical studies may be accessible to the public.

2.4.7 Medical Images and Diagnosis

In the medical field, AI is being used to speed up the process of administering personalized medications to patients, increase the accuracy of prognoses, and enhance the quality of diagnostics. The application of AI is causing profound shifts in the medical imaging and diagnostics industries. Deep learning algorithms have been introduced, which can rapidly and precisely detect abnormalities in medical images. This could potentially reduce the need for invasive procedures. A recent study showed that deep learning can accurately diagnose breast cancer when applied to mammography scans [36].

Diagnosis is another area where AI is being utilized to better the care and treatment of patients. To better forecast patient outcomes, for instance, machine learning algorithms have been developed, which use patient data to suggest individualized treatment strategies. Another study used machine learning to forecast the likelihood of cardiac failure in diabetic individuals [37].

Numerous studies have pointed out AI's potential to enhance the precision of diagnoses, the reliability of prognostications, and the flexibility of treatment options for individual patients. Here are some specific examples.

1. Algorithms powered by AI can detect cancer more accurately in medical imaging compared to human radiologists. In one study [38], AI algorithm was able to predict the risk of cardiac attack and the accuracy level of the detection was 85%.
2. AI is also being used to figure out who is most likely to gain from preventative treatments for any sudden heart diseases and stroke as well [39] and another study which is published in the journal *Cancer Cell* found that an AI algorithm was able to forecast which cancer patients with 85% precision would benefit from immunotherapy [40].

2.4.8 Role of Blockchain and AI in Healthcare

AI significantly revolutionizes the healthcare business by offering improved data analysis, decision assistance, and predictive modeling. In one study [41], the researchers highlighted the importance of deep learning methods in healthcare such as diagnosing disease from medical images, with the help of NLP extraction of information in clinical notes and Analysis of past patient data to create individualized care plans. For better understanding and better diagnoses of patients and planning their treatments, AI algorithms have the potential to examine massive amounts of medical data.

For several problems in the healthcare industry, including the need for reliable and consistent data exchange, blockchain technology has emerged as a promising option. blockchain technology for medical records and research data is examined in this case study of the "MedRec" prototype [42]. Blockchain provides a distributed and immutable ledger, improving stored information's confidentiality, integrity, and verifiability. This method ensures that patients consent and that data is kept intact while being shared across various parties involved in their care.

Undoubtedly, AI and blockchain have enormous potential to revolutionize the healthcare industry by strengthening data security, medical record keeping, and drug supply chain management (Figure 2.4).

AI algorithms can analyze and interpret complex healthcare data, helping with early disease diagnosis, therapy planning for individuals and monitoring patients. Blockchain technology can improve data privacy, integrity, and interoperability while offering a decentralized and unchangeable infrastructure. Better patient outcomes, reduced costs, and increased efficiency might result from applying AI to blockchain technology in the healthcare industry. Data security, scalability, regulatory frameworks, and standardization are just some issues that must be resolved before the healthcare industry can fully benefit from AI and blockchain.

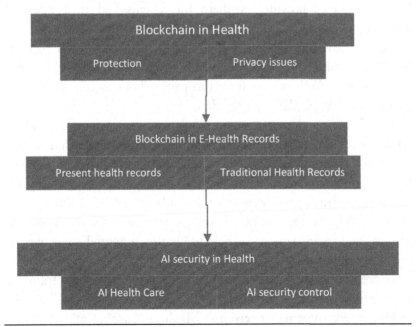

Figure 2.4 AI and blockchain in healthcare.

2.5 Enhancing the Healthcare Sector with AI and Blockchain Technology

The integration of AI and blockchain technology can create huge numbers of applications in the healthcare sector. There is a growing trend of using smart devices to monitor patients' health and tracking. As a result, this is creating a great deal of data which numerous AI-based systems may use for analysis and determination. When much data are involved, providing privacy and security is the main concern. Here we will talk about the privacy of the patients, the importance of medical records keeping with the help of AI and blockchain technology and the involvement in medicine supply chain administration.

2.5.1 Patient Privacy

Privacy is an important concern in many sectors today. People's privacy is more at the forefront of their minds as technology advances since certain services require access to sensitive data. AI and blockchain technology can strengthen patient data security by ensuring vital medical records' privacy, authenticity, and availability. Blockchain's immutability and AI data analysis might result in robust protection. More than 80 million health insurance provider Anthem Inc. clients had their personal information compromised in a 2015 cyberattack [43]. Records containing sensitive information, such as names, birthdays, Social Security Numbers, and Healthcare Identification Numbers (HICNs), were compromised. Anthem was forced to pay millions to consumers in a class action lawsuit filed after the data leak.

So, protecting data esp. patients' data is very important and in 2016 few researchers [44] established their assumptions based on one case study called "MedRec" where they emphasized that blockchain technology can provide data security, privacy, and auditability by enabling security, privacy, and audit trails for all data accesses. In terms of AI, AI can analyze large amounts of data to categorize patterns and trends which can help to identify potential security risks and vulnerabilities [45].

Therefore, today's challenge is determining how best to train AI to recognize patterns and insights in patient data that will aid in making more informed decisions. Machine learning allows AI to analyze patient data for trends and insights. Algorithms make feasible what would be exceedingly difficult or impossible to accomplish without them: sifting through huge amounts of data to find patterns and trends.

Machine learning has been explored in a work published in the *Journal of Medical Internet Research* to better tailor treatments to individual patients and foresee the success or failure of medical procedures. In that study, researchers analyzed electronic medical data using machine learning algorithms to determine the characteristics associated with unfavorable outcomes [46].

In another study published in the journal *Nature*, medicine found that machine learning algorithms can analyze medical images and identify patterns that relate to some specific disease. In that study, Machine learning was employed in the research to analyze photos of skin lesions and detect patterns linked with melanoma [47].

2.5.2 Medical Record Keeping

With the potential to improve data interoperability, accessibility, and integrity, AI and blockchain hold great promise for the future of medical record keeping. Here are several ways that medical record keeping might be improved with the use of AI and blockchain technology, along with some reading recommendations and examples of modern applications:

1. **Automating Administrative Tasks:** To reduce time and error, AI can automate administrative tasks related to medical record keeping [48].
2. **Improving Accuracy and Completeness:** By flagging any incomplete or inaccurate entries, AI can ensure patients' medical files are comprehensive and reliable [49].
3. **Enhancing Data Analytics:** By examining patient information, AI can assist medical professionals in recognizing patterns and trends that might direct their treatment [47].
4. **Ensuring Privacy and Security:** Patients' privacy and the medical records' accuracy can be protected by blockchain technology. One way to achieve this goal is to encrypt patient data before distributing it over several nodes [44].

One notable example of the use of AI in medical record keeping is the partnership between Google and Ascension, a major healthcare provider in the United States. Google uses AI to sift through Ascension's medical information in search of patterns that might inform future care decisions [50].

Analyzing medical records can benefit from using NLP as both a framework and a tool. NLP is the language comprehension subfield of AI. NLP allows for analyzing medical records to discover patterns and trends that would be impossible to discover manually [51].

The **Medicalchain** platform uses distributed ledger technology (blockchain) and AI. Patients are afforded the right to retain their medical records, control who has access to those data, and obtain aid in identifying and fixing any errors that may have been made in those records [48]. Improved patient outcomes at lower systemic costs are expected because of **Medicalchain's** use of AI in tandem with blockchain technology [52].

2.5.3 Drug Supply Chain Administration

By increasing the transparency, authenticity, and traceability of pharmaceutical items, AI and Blockchain have the potential to significantly improve the management of the medication supply chain. They can improve drug supply chain management and authenticity in several ways:

1. **Tracking the Movement of Drugs:** Using blockchain technology, the distribution chain of drugs may be tracked from the manufacturer to the end user. This is useful for ensuring that medicines have not been altered or are genuine [44].
2. **Ensuring Authenticity:** By providing an immutable record of the drug's origin and distribution, blockchain technology can be implemented to ensure that the pharmaceuticals being sold are genuine [53].
3. **Predicting Drug Demand:** The need for medications can be anticipated using AI by analyzing historical data and current factors such as shifting populations and disease outbreaks. This can help ensure that individuals have access to medications at the times when they are required to take them (Figure 2.5) [54].

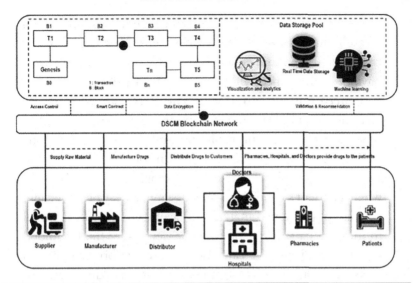

Figure 2.5   The overview of the drug supply chain by the use of blockchain technology [55].

The collaboration between IBM and Walmart proved how blockchain technology can help pharmaceutical companies coordinate the distribution of their medicine. Their joint effort with the help of blockchain technology helps them to monitor the movement of food from farm to factory. It's also possible for all pharmaceutical companies to benefit from that technology [56].

To protect the confidentiality and accessibility of patient data, the Estonian government has approved the use of blockchain technology in the country's healthcare system. eHealth Estonia aims to give people more control over their medical records while ensuring confidentiality and making that data easily accessible to healthcare providers.

2.6 The Potential of AI and Blockchain to Improve Medical Care

AI and blockchain technology could make a big difference in medical care by making it easier to share and use data and make decisions.

1. **Improve Diagnosis and Treatment:** As we have already discussed, AI can collect large amounts of data to identify patterns and trends which can help doctors diagnose diseases more accurately.For example, AI algorithms can detect skin cancer more accurately than human dermatologists [57].

Using a patient's medical history and genetic makeup, AI can create individualized treatment regimens. New cancer medications that are more effective and have fewer adverse effects have been developed with the help of AI [58].

2. **Enhanced Patient Monitoring and Personalized Care:** By analyzing data in real time from wearable devices, EHRs, and other sources, AI can facilitate continuous patient monitoring and customized therapy. In one study [59], AI systems can analyze patient physiological data to predict when a patient's condition will worsen in a healthcare setting. The use of predictive analytics in healthcare has the potential to improve patient outcomes while decreasing the likelihood of adverse effects. AI can also help to perform automated tasks such as scheduling appointments, making patients' records, and processing insurance claims, freeing healthcare professionals to spend more time on patient care [60].

3. **Secure and Interoperable Data Sharing:** Blockchain technology provides a decentralized and unchangeable platform when sharing and exchanging sensitive information. This can help make it easier for doctors to share patient information, benefiting patient care [44].

4. **Drug Discovery Process:** AI can assist in speeding up the discovery of new medicines by identifying promising new drug candidates and facilitating their rapid evaluation for safety and efficacy. Applying AI has led to new treatments for diseases, including Alzheimer's and cancer [34].

5. **Clinical Trials:** Clinical trials can become more transparent and safer when blockchain technology is utilized. TrialX is a firm using blockchain technology to provide a decentralized platform for clinical trials [61]. On this platform, patients can maintain ownership of their data, and the results of the clinical studies may be accessible to the public.

6. **Fraud Prevention:** Blockchain can be securely used for storing and sharing information which can help organizations to prevent fraud. For instance, a business known as Gem is utilizing blockchain technology to provide a decentralized platform for processing medical claims [62]. This platform will lower the likelihood of fraudulent activity and

errors. SimplyVital Health, a second firm, is utilizing blockchain technology to build a decentralized network to share healthcare data. This platform will increase data security as well as data privacy.

According to the *Journal of Medical Internet Research*, blockchain and AI might enhance China's EHR administration [59]. According to the study [59], blockchain can improve EHR security and privacy, while AI can increase diagnostic and treatment accuracy and efficiency. The study examined a Chinese hospital's blockchain-based EHR system with AI. The system used AI to evaluate patient data and provide personalized treatment suggestions while securely storing and sharing it with healthcare practitioners. The technology also monitored EHR usage and prevented unauthorized access. Research shows that blockchain and AI increased diagnostic and treatment accuracy and efficiency. The system correctly diagnosed a breast cancer patient based on her medical history and imaging data. The system also suggested personalized therapy choices based on the patient's needs. According to the report, blockchain and AI might transform healthcare in China and elsewhere. It highlighted that the technology could increase patient data security and diagnostic and treatment accuracy.

2.7 Case Studies

In this chapter, we'll examine two case studies to learn more about the processes involved in implementing these technologies and the outcomes those initiatives yielded for their respective companies. After this talk, we will record the good and negative results.

2.7.1 The Pfizer–IBM MedsChain

Pfizer, a pharmaceutical powerhouse, and IBM, a technology giant, worked together on an initiative to improve the pharmaceutical supply chain using blockchain and AI. The logistics of the production of pharmaceuticals are well-documented. Each link in the medication supply chain—from producers to distributors to retailers to pharmacies to consumers—is crucial. Because of the complexity of this method, ensuring the safe and effective delivery of medications is challenging. Pfizer and IBM collaborated to create the blockchain-based **MedsChain** to address this issue in the pharmaceutical supply chain. This **MedsChain** was created to lower healthcare costs, increase transparency, and enhance patient outcomes. **MedsChain** utilizes blockchain technology to construct a trustable, decentralized record of pharmaceutical transactions that only authorized parties can modify along the supply chain. By creating an immutable record of encrypted drug transactions that authorized parties can only view, the system helps reduce the prevalence of counterfeit drugs and improves patient safety.

2.7.1.1 Example   Suppose the patient's name is Sarah who needs a prescription medicine to treat her condition. A pharmaceutical business manufactures the drug, then distributes it to a pharmacy where Sarah may pick it up. Each step in the medication supply chain, from manufacturing to delivery, would be documented on a blockchain ledger using **MedsChain** technology. This would generate an immutable record of each transaction, guaranteeing that the drug is genuine and unaltered.

After picking up the prescription at the pharmacy, Sarah may get a notification on her phone or other device asking her to fill out a form with details about her current and past medical conditions and medications. This information would be transmitted securely to an AI system to analyze and spot drug-to-drug interactions or negative side effects. Sarah's phone, pharmacist, or healthcare provider may receive a warning from the AI system if problems are detected, prompting Her to take the necessary steps. One medical expert could recommend more monitoring or testing, while another would recommend a different medication or dosage.

By combining AI with blockchain technology, the MedsChain system has the potential to vastly enhance the medication supply chain, resulting in better patient outcomes and enhanced healthcare delivery.

2.7.1.2 Case Study 2: Deephealth Project   German institutions, the Technical University of Munich and the University of Lübeck spearheaded DeepHealth. The EU funded this study as part of its Horizon 2020 initiative to enhance healthcare through research and innovation [63].

Diabetic retinopathy is a frequent complication of diabetes that can cause blindness if left untreated. Thus the DeepHealth project developed an AI-powered platform to analyze medical imaging data to detect it. The system has a sensitivity of 96.1% and a specificity of 93.6% for detecting diabetic retinopathy from 10,000 images of diabetic patients' retinas [64].

Let's assume a diabetic visits an eye doctor for an annual exam. The ophthalmologist uses a retina camera to capture an image of the back of the patient's eye. A crystal-clear picture of the retina is seen here. The idea is then uploaded to a service where AI examines it. The platform's deep learning algorithms analyze the photo for symptoms of diabetic retinopathy.

The platform may explain the diagnosis by, for instance, highlighting areas of the picture that exhibit evidence of diabetic retinopathy if the illness is detected. The ophthalmologist can examine the platform's Analysis of the patient's condition to discover the cause of the problem and the best course of treatment.

2.8 A Comprehensive Assessment of the Positive and Negative Aspects of Implementation

AI and blockchain technology are neat tools. To keep patient data secure and predict illnesses before they happen, these two technologies can help. However, there are some drawbacks and positive impacts of this technology.

2.8.1 Positive Impact

1. **Improved Data Security and Privacy:** To lessen the likelihood of data breaches and unauthorized access, patient information might be stored and shared via a blockchain-based, decentralized platform [44,65].
2. **Streamlined Administrative Processes:** Administrative tasks, including billing and claims processing, may be simplified and automated using blockchain technology, leading to fewer mistakes and greater productivity [66].
3. **Personalized Medicine:** Through the Analysis of patient data, AI can enable personalized medicine by allowing

therapies to be adapted to the individual. The outcomes and costs of medical treatment may benefit from this [67].

4. **Enhanced Predictive Analysis:** AI makes the capacity to intervene early in a disease's progression possible [68].

2.8.2 Negative Impact

1. **Workplace Displacement:** Some healthcare workers, especially those in administrative positions, may lose their jobs because of the widespread use of AI and blockchain technology in the healthcare industry [69].
2. **Technical Limitations:** Implementing AI and blockchain in healthcare may be challenging due to technological challenges such as scale and the necessity for standardized data formats [70].
3. **Scalability and Performance:** Using AI in conjunction with blockchain technology in healthcare might be challenging due to limitations in size and speed. Blockchain networks' data processing speeds and accuracy are very variable depending on the consensus mechanisms employed. AI algorithms that require near-instantaneous responses might be slowed down as a result [71].

2.9 Open Challenges for AI and Blockchain Technologies for the Healthcare Sector

Benefits do not come along without a cost. While the transformational effects of blockchain and AI technologies in the healthcare sector are undeniable, these advancements have not come without a price tag. These technologies had a hefty initial investment and required substantial local data storage. Below are some statistics that provide a bird's eye view of the problems that AI and Blockchain technology are now facing in the medical field.

2.9.1 Privacy Issues

Protecting patients' privacy is very important since their personal information is stored in healthcare databases. Privacy concerns arise when people share data on public blockchains. Because blockchain is decentralized and transparent, data protection has big risks. Different approaches, such as distributed consensus, have been established to deal with the privacy concerns highlighted by blockchain technology [72].

2.9.2 *Resource Limitation, Capacity, and Storage of Data*

Blockchain and AI demand various data storage and capacity. Recent blockchain technology has a cap on the number of simultaneous network transactions. That means high consumption of power, which also leads to increased energy consumption, is required to process the transaction [72]. In contrast, AI-based systems require sophisticated hardware. Not only the hardware, but also it involves complex calculations with many algorithms that process huge amounts of data in every cycle (Figure 2.6).

2.9.3 *Blockchain Security and Threats*

The weaknesses in blockchain's security have been recently exposed in research. Attackers can readily uncover protocol problems in

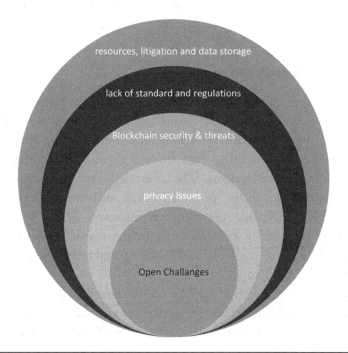

Figure 2.6 Open challenges.

blockchain systems because they haven't improved much in recent years. Several potential risks are associated with blockchain technology, such as data tampering and theft. AI learning and model output can be impacted by malicious attackers injecting false data or sample inputs, which poses a risk to the health and safety of consumers. It has been discovered that blockchain technology is vulnerable to cyberattacks such as the DNS (Domain Name System) attack and mempool attacks [29].

2.9.4 Lack of Standards, Interoperability, and Regulations

As we already know, there should be regulatory laws before implementing anything. Most AI and blockchain integration is still limited to academic research, and few applications have made it possible. One of the main challenges with adopting blockchain in healthcare is the question of who owns what. Since both public and private blockchains exist, one would reasonably wonder who is legally responsible for the information stored there. Many legal and regulatory problems arise when there is no central body to answer to for the consequences of technological action [29].

2.10 Future Research Direction

The model (AI and blockchain technology) requires more investigation because the field is rapidly evolving. The privacy and security of blockchain with AI still have a long way to go. Future studies can also focus on implementing modern hash algorithms and employing secure authorization systems. The advantages and disadvantages of implementing two-factor or multi-factor authentication at every node in the blockchain network are unknown. Many researchers have detected security vulnerabilities in the classic consensus protocol, such as Proof-of-Work and Proof-of-Stake, and suggested not to use the protocol anymore (Figure 2.7).

To further improve the stability and efficiency of the system, this new AI-powered blockchain may adopt unique consensus protocols like proof-of-elapsed-time and Delegated Proof-of-Stake. Compared to a non-cooperative computing technique, cooperative computing is believed to shorten the time required to calculate a block's hash [72].

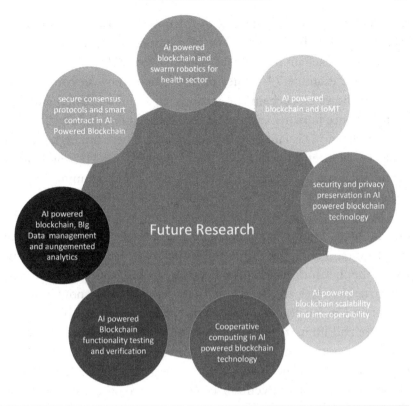

Figure 2.7 Future research of the integration of AI and blockchain technology.

Future studies could investigate the feasibility of merging AI-powered blockchain with preexisting systems and boosting the system's processing speed. It is hypothesized [73] that the time needed to calculate a block's hash can be decreased by using cooperative computing instead of non-cooperative computing. Future studies may also look toward integrating blockchain with AI-powered current systems. If blockchain systems driven by AI are going to be used in the healthcare industry on a commercial scale, security and privacy must be considered first.

2.11 Conclusion

This section provides an overview of how blockchain and AI are applied in medicine. A lot of study has been done during the COVID-19 era on the practical implications of these two. Many scholars have proposed changes to existing models, and some have even suggested the development of a futuristic representation. In 2022, the blockchain market was valued at over $500 million. In this chapter, we examine 74 studies on the issue of how to combine AI with healthcare blockchain applications. In addition, the potential of AI to enhance humans' capacity to recognize and respond swiftly to a medical diagnosis is laid bare in this chapter. There has been a lot of study on how effectively blockchain can protect EHRs and other confidential medical records. Data processing and scanning are two tasks that may be automated by combining AI with ML.

The chapter has provided a high-level summary of the functioning of the blockchain system and the many layers included in the architecture. The application and presentation layer, the consensus layer, the network layer, the data layer, and the hardware and infrastructure layer are the five primary layers that are now in place. Despite the many developments in machine learning and blockchain, there are still limits to the present system. While the security of the blockchain system's networked data storage is undeniable, users have legitimate privacy concerns with this cutting-edge technology. Public blockchain databases, decentralized systems, and other aspects can be compromised by outsiders, giving hackers easy access to the underlying data. There are also regulatory constraints on the use of blockchain and AI technology on medical databases, which is a major roadblock to the development of AI-powered blockchain. Future research prospects addressing these constraints and difficulties have also been covered.

References

[1] Blockchain: Foundations and use cases. *My Mooc. [Online]*. Available: https://www.my-mooc.com/en/mooc/blockchain-foundations-and-use-cases/. [Accessed: 05-Jun-2023].

[2] Meyer, M. Blockchain technology: Principles and applications. *IEEE Access*, 13-Feb-2020. [Online]. Available: https://ieeeaccess.ieee.org/closed-special-sections/blockchain-technology-principles-and-applications/. [Accessed: 05-Jun-2023].

[3] Guru, A., Mohanta, B. K., Mohapatra, H., Al-Turjman, F., Altrjman, C., & Yadav, A. (2023). A survey on consensus protocols and attacks on blockchain technology. *Applied Sciences (Basel)*, 13(4), 2604.

[4] Wang, Y. (2021). The challenges and opportunities of blockchain systems. In *2021 2nd International Conference on Computer Communication and Network Security (CCNS)*, pp. 84–88.

[5] Dorri, A., Steger, M., Kanhere, S. S., & Jurdak, R. (2017). BlockChain: A distributed solution to automotive security and privacy. *arXiv [cs.CR]*.

[6] Agbo, C. C., Mahmoud, Q. H., & Eklund, J. M. (2019). Blockchain technology in healthcare: A systematic review. *Healthcare*, 7(2), 56.

[7] Mehta, R., & Dubey, A. K. (2018). Blockchain and healthcare: A comprehensive survey. *International Journal of Research in Electronics and Computer Engineering*, 5(3), 647–652.

[8] Russell, S. J., & Norvig, P. (2010). *Artificial intelligence: A modern approach*. Prentice Hall Press.

[9] Azaria, A., Ekblaw, A., Vieira, T., & Lippman, A. (2016). Medrec: Using blockchain for medical data access and permission management. In Proceedings of the *2016 2nd International Conference on Open and Big Data (OBD)* (pp. 25–30). IEEE.

[10] Li, M., Jiang, P., Li, Y., & Chen, W. (2019). Blockchain-based medical records secure storage and medical service framework. *Journal of Medical Internet Research*, 21(2), e12992. 10.2196/12992

[11] Abdullah, R., Bahsoon, R., & Ji, Y. (2019). A systematic review of blockchain-based secure and privacy-preserving electronic health record system. *IEEE Access*, 7, 153930–153941.

[12] Wang, S., et al. (2019). Artificial intelligence in healthcare: Past, present and future. *American Journal of Managed Care*, 25(12), e393–e398.

[13] Cresswell, K. M., & Sheikh, A. (2011). Organizational issues in the implementation and adoption of health information technology innovations: An interpretative review. *International Journal of Medical Informatics*, 80(5), 73–83.

[14] Li, J., Zhang, Y., & Chen, Y. (2020). Blockchain-based secure healthcare systems: A systematic review. *Journal of Medical Systems*, 44(3), 1–12.

[15] FDA (2019). DSCSA implementation: Product tracing requirements for dispensers - compliance policy. *U.S. Food and Drug Administration*.

[16] Dong, X., et al. (2019). Blockchain-based patient-controlled medical record sharing system. *Journal of Medical Systems*, 43(7), 1–9.

[17] Market Research Future (2019). Blockchain in healthcare market research report - global forecast till 2023. *Market Research Future*.

[18] McKinsey & Company. (2020). Artificial Intelligence: Healthcare's new nervous system. *McKinsey*.

[19] Bell, L., Buchanan, W. J., Cameron, J., & Lo, O. (2018). Applications of blockchain within healthcare. *Blockchain in Healthcare Today*, 1. 10.30953/bhty.v1.8

[20] Ghosh, A. & Mistri, B. (2020). Spatial disparities in the provision of rural health facilities: Application of GIS based modelling in rural Birbhum, India. *Spatial Information Research*, 28(6), 655–668.

[21] McGhin, T., Choo, K.-K. R., Liu, C. Z., & He, D. (2019). Blockchain in healthcare applications: Research challenges and opportunities. *Journal of Network and Computer Applications*, 135, 62–75.

[22] Awad Abdellatif, A. et al. (2021). MEdge-chain: Leveraging edge computing and blockchain for efficient medical data exchange. *IEEE Internet of Things Journal*, 8(21), 15762–15775.

[23] Shahnaz, A., Qamar, U., & Khalid, A. (2019). Using blockchain for electronic health records. *IEEE Access*, 7, 147782–147795.

[24] Vora, J. et al. (2018). BHEEM: A blockchain-based framework for securing electronic health records. In *2018 IEEE Globecom Workshops (GC Wkshps)*.

[25] J. O. Healthcare Engineering (2023). Retracted: Research on the application of blockchain in smart healthcare: Constructing a hierarchical framework. *Journal of Healthcare Engineering*, 2023, 9850894.

[26] Katuwal, G. J., Pandey, S., Hennessey, M., & Lamichhane, B. (2018). Applications of blockchain in healthcare: Current landscape & challenges. *arXiv [cs.CY]*.

[27] Subramanian, G. & Sreekantan Thampy, A. (2021). Implementation of blockchain consortium to prioritize diabetes patients' healthcare in pandemic situations. *IEEE Access*, 9, 162459–162475.

[28] Ahmad, R. W., Salah, K., Jayaraman, R., Yaqoob, I., Ellahham, S., & Omar, M. (2021). The role of block chain technology in telehealth and telemedicine. *International Journal of Medical Informatics*, 148(104399), 104399.

[29] Nguyen, D. C., Ding, M., Pathirana, P. N., & Seneviratne, A. (2021). Blockchain and AI-based solutions to combat Coronavirus (COVID-19)-like epidemics: A survey. *IEEE Access*, 9, 95730–95753.

[30] Chen, H., Engkvist, O., Wang, Y., Olivecrona, M., & Blaschke, T. (2018). The rise of deep learning in drug discovery. *Drug Discovery Today*, 23(6), 1241–1250.

[31] Zhang, Q., Zhu, L., & Huang, D.-S. (2019). High-order convolutional neural network architecture for predicting DNA-protein binding sites. *IEEE/ACM Transactions on Computational Biology and Bioinformatics*, 16(4), 1184–1192.

[32] Li, M. et al. (2019). Automated ICD-9 coding via a deep learning approach. *IEEE/ACM Transactions on Computational Biology and Bioinformatics*, 16(4), 1193–1202.

[33] Liu, B. & Li, S. (2019). ProtDet-CCH: Protein remote homology detection by combining long short-term memory and ranking methods. *IEEE/ACM Transactions on Computational Biology and Bioinformatics*, 16(4), 1203–1210.

[34] Yu, K.-H., Beam, A. L., & Kohane, I. S. (2018). Artificial intelligence in healthcare. *Nature Biomedical Engineering*, 2(10), 719–731.

[35] Patel, C., Gomadam, K., Khan, S., & Garg, V. (2010). TrialX: Using semantic technologies to match patients to relevant clinical trials based on their personal health records, *Web Semantic*, 8(4), 342–347.

[36] Liu, S. et al. (2019). Deep learning in medical ultrasound analysis: A review. *Engineering (Beijing)*, 5(2), 261–275.

[37] Segar, M. W. et al. (2019). Machine learning to predict the risk of incident heart failure hospitalization among patients with diabetes: The WATCH-DM risk score. *Diabetes Care*, 42(12), 2298–2306.

[38] McKinney, S. M. et al. (2020). International evaluation of an AI system for breast cancer screening. *Nature*, 577(7788), 89–94.

[39] Deberneh, H. M. & Kim, I. (2021). Prediction of type 2 diabetes based on machine learning algorithm. *International Journal of Environmental Research and Public Health*, 18(6), 3317.

[40] Xu, Z., Wang, X., Zeng, S., Ren, X., Yan, Y., & Gong, Z. (2021). Applying artificial intelligence for cancer immunotherapy. *Acta Pharmaceutica Sinica B*, 11(11), 3393–3405.

[41] Miotto, R., Wang, F., Wang, S., Jiang, X., & Dudley, J. T. (2018). Deep learning for healthcare: Review, opportunities and challenges. *Briefings in Bioinformatics*, 19(6), 1236–1246.

[42] Azaria, A. A case study for blockchain in healthcare: "MedRec" prototype for electronic health records and medical research data. *MIT Media Lab. [Online]*. Available: https://www.media.mit.edu/publications/medrec-whitepaper/. [Accessed: 03-Jun-2023].

[43] Reuters (05-Feb-2015). Massive Anthem health insurance hack exposes millions of customers' details. *The Guardian*.

[44] Tarallo, E., Akabane, G. K., Shimabukuro, C. I., Mello, J., & Amancio, D. (2019). Machine learning in predicting demand for fast-moving consumer goods: An exploratory research. *IFAC-PapersOnLine*, 52(13), 737–742.

[45] Vyas, S., Shabaz, M., Pandit, P., Parvathy, L. R., & Ofori, I. (2022). Integration of artificial intelligence and blockchain technology in healthcare and agriculture. *Journal of Food Quality*, 2022, 1–11.

[46] Topol, E. J. (2019). High-performance medicine: The convergence of human and artificial intelligence. *Nature Medicine*, 25(1), 44–56.

[47] Rajkomar, A., Dean, J., & Kohane, I. (2019). Machine learning in medicine. *The New England Journal of Medicine*, 380(14), 1347–1358.

[48] Weng, S. F., Reps, J., Kai, J., Garibaldi, J. M., & Qureshi, N. (2017). Can machine-learning improve cardiovascular risk prediction using routine clinical data?. *PLoS One*, 12(4), e0174944.

[49] Sidey-Gibbons, J. A. M. & Sidey-Gibbons, C. J. (2019). Machine learning in medicine: A practical introduction. *BMC Medical Research Methodology*, 19(1), 64.

[50] Pilkington, E. (12-Nov-2019). Google's secret cache of medical data includes names and full details of millions – whistleblower. *The Guardian*.

[51] Wang, Y., Tafti, A., Sohn, S., & Zhang, R. (2019). Applications of natural language processing in clinical research and practice. In *Proceedings of the 2019 Conference of the North*, pp. 22–25.

[52] Us, A. Home. *Medicalchain. [Online]*. Available: https://medicalchain. com/en/. [Accessed: 04-Jun-2023].

[53] Badhotiya, G. K., Sharma, V. P., Prakash, S., Kalluri, V., & Singh, R. (2021). Investigation and assessment of blockchain technology adoption in the pharmaceutical supply chain. *Materials Today*, 46, 10776–10780.

[54] Walmart case study – hyperledger foundation. *Hyperledger.org*. 11-Jan-2019. [Online]. Available: https://www.hyperledger.org/learn/ publications/walmart-case-study. [Accessed: 04-Jun-2023].

[55] Abbas, K., Afaq, M., Ahmed Khan, T., & Song, W.-C. (2020). A blockchain and machine learning-based drug supply chain management and recommendation system for smart pharmaceutical industry. *Electronics (Basel)*, 9(5), 852.

[56] Esteva, A. et al. (2017). Dermatologist-level classification of skin cancer with deep neural networks. *Nature*, 542(7639), 115–118.

[57] Tagde, P. et al. (2021). Blockchain and artificial intelligence technology in e-Health. *Environmental Science and Pollution Research*, 28(38), 52810–52831.

[58] Buch, V. H., Ahmed, I., & Maruthappu, M. (2018). Artificial intelligence in medicine: current trends and future possibilities. *British Journal of General Practice*, 68(668), 143–144.

[59] Rajkomar, A. et al. (2018). Scalable and accurate deep learning with electronic health records. *NPJ Digital Medicine*, 1(1), 18.

[60] TrialX. *Trialx.com. [Online]*. Available: https://trialx.com/. [Accessed: 04-Jun-2023].

[61] Finance, Digital, Blockchain, and Innovation. Gem: Why we're building the blockchain for healthcare. *Corporate Finance, DeFi, Blockchain News. [Online]*. Available: https://www.finyear.com/ Gem-Why-We-re-Building-the-Blockchain-for-Healthcare_ a36058.html. [Accessed: 04-Jun-2023].

[62] Hazlegreaves, S. (01-Oct-2019). DeepHealth project: Deep-learning and HPC to boost biomedical applications for health. *Open Access Government. [Online]*. Available: https://www.openaccessgovernment.org/biomedical-applications-for-health/74475/. [Accessed: 04-Jun-2023].

[63] Kwak, G. H. & Hui, P. (2019). DeepHealth: Review and challenges of artificial intelligence in health informatics. *arXiv [cs.LG]*.

[64] Revolutionizing healthcare with blockchain and artificial intelligence. *Cbcamerica.org. [Online]*. Available: https://www.cbcamerica.org/ blockchain-insights/revolutionizing-healthcare-with-blockchain-and-artificial-intelligence. [Accessed: 05-Jun-2023].

[65] Benchoufi, M., Porcher, R., & Ravaud, P. (2018). Blockchain protocols in clinical trials: Transparency and traceability of consent. *F1000Res.*, 6, 66.

[66] Ichikawa, D., Kashiyama, M., & Ueno, T. (2017). Tamper-resistant mobile health using blockchain technology. *JMIR MHealth UHealth*, 5(7), e111.

[67] Esteva, A. et al. (2019). A guide to deep learning in healthcare. *Nature Medicine*, 25(1), 24–29.

[68] Aloini, D., Benevento, E., Stefanini, A., & Zerbino, P. (2023). Transforming healthcare ecosystems through blockchain: Opportunities and capabilities for business process innovation. *Technovation*, 119(102557), 102557.

[69] Meskó, B., Drobni, Z., Bényei, É., Gergely, B., & Győrffy, Z. (2017). Digital health is a cultural transformation of traditional healthcare. *MHealth*, 3, 38.

[70] Yue, X., Wang, H., Jin, D., Li, M., & Jiang, W. (2016). Healthcare data gateways: Found healthcare intelligence on blockchain with novel privacy risk control. *Journal of Medical Systems*, 40(10), 218.

[71] Mazlan, A. A., Mohd Daud, S., Mohd Sam, S., Abas, H., Abdul Rasid, S. Z., & Yusof, M. F. (2020). Scalability challenges in healthcare blockchain system—A systematic review. *IEEE Access*, 8, 23663–23673.

[72] Fusco, A., Dicuonzo, G., Dell'Atti, V., & Tatullo, M. (2020). Blockchain in healthcare: Insights on COVID-19. *International Journal of Environmental Research and Public Health*, 17(19), 7167.

[73] Wu, D. & Ansari, N. (2020). A cooperative computing strategy for blockchain-secured fog computing. *IEEE Internet of Things Journal*, 7(7), 6603–6609.

3

Developing Trading Strategies in Decentralized Market Prediction by Using AI, ML, and Blockchain Technology

MD ABU SUFIAN

3.1 Introduction

3.1.1 Background and Context

Blockchain technology and artificial intelligence (AI) are two of the most revolutionary and disruptive technologies nowadays. Both technologies have the potential things to transform many industries, and in some cases, they already have essential features, uses, obstacles, and opportunities.

3.1.2 Problem Statement

The potential of AI, machine learning (ML), and blockchain technology in decentralized trading markets is yet to be fully explored and understood. Their implementation raises critical questions about market prediction accuracy, security enhancements, ethical considerations, and regulatory compliance. This necessitates an urgent and comprehensive investigation into the use and implications of these disruptive technologies in the context of decentralized trading markets.

DOI: 10.1201/9781003162018-3

3.1.3 Artificial Intelligence

AI describes the creation of computer systems that can carry out tasks that would typically require human intelligence, such as comprehending natural language, recognizing patterns in images, making judgement calls, and picking up new skills over time. Two types of AI exist strong or general AI, which can perform any intellectual task that a human is capable of, and narrow or weak AI, which is limited to performing specific tasks within a narrow domain. Numerous industries, including healthcare, finance, transportation, and education, have adopted AI in a variety of ways. AI can be used in healthcare to diagnose illnesses, track patient health, and create individualized treatment plans [1]. AI in finance can be applied to automate financial processes, analyse market trends, and detect fraud [2]. AI can be applied to the development of self-driving cars and the optimization of logistics [3]. AI can be used in education to provide intelligent tutoring, assess student progress, and personalize learning [4]. But AI also brings with it several difficulties, such as ethical issues like the significance of bias and discrimination, the absence of transparency and accountability, and the potential for job displacement [5]. Researchers and decision-makers are working to create ethical standards, legal frameworks, and educational initiatives to address these issues [6].

3.1.4 Blockchain Technology

A distributed ledger made possible by blockchain technology allows for secure, transparent transactions between many participants without the necessity of middlemen. 2008 saw the introduction of the underlying technology behind Bitcoin, the first decentralized digital money. Among the many uses that blockchain technology has since advanced to enable our supply chain management, digital identity verification, smart contracts, and decentralized finance. Supply chain management may use blockchain technology to track product provenance and authenticity, reducing fraud and counterfeiting while increasing efficiency and transparency [7]. To generate safe, decentralized digital identities that can be confirmed without the need for a centralized authority, blockchain technology can be employed in digital identity verification [8]. Smart contracts can employ blockchain technology to automate the execution of contracts and ensure compliance [9]. Decentralized financial systems that are more accessible, safe, and transparent than traditional financial systems can be created using blockchain technology [10]. Scalability, interoperability, and regulatory uncertainty are some of the issues that blockchain technology faces [11]. Researchers and developers are working on a variety of solutions, including scaling solutions, interoperability protocols, and regulatory frameworks [12,13], to address these challenges.

3.1.5 Aim

The aim of this research is to develop and evaluate the model performances on trading strategies-based AI, ML, and blockchain technology in decentralized markets.

3.1.6 Research Questions

1. How effective are AI and ML techniques in predicting market trends and generating profitable trading signals in decentralized trading markets?
2. How can blockchain technology be used to enhance the transparency, security, and efficiency of trading in decentralized trading markets?
3. What types of trading indicators are most reliable for predicting market movements in decentralized trading markets, and how can they be incorporated into AI and ML models?
4. What ethical and regulatory issues arise when incorporating AI and ML models into decentralized trading markets, and how can these be addressed?
5. How do decentralized trading markets compare to centralized trading markets in terms of efficiency, transparency, and security, and what are the implications for investors and traders?

3.1.7 Hypothesis

3.1.7.1 Hypothesis 1 ML and AI techniques can accurately predict market trends and generate profitable trading signals in decentralized trading markets.

3.1.7.2 Hypothesis 2 The implementation of blockchain technology can improve the transparency, security, and efficiency of trading in decentralized trading markets.

3.1.7.3 Hypothesis 3 Integrating multiple trading and tweeter stock indicators into ML and AI models can enhance the precision of market predictions in decentralized trading markets.

3.1.7.4 Hypothesis 4 The incorporation of AI and ML models into decentralized trading markets must consider ethical and regulatory concerns.

3.1.8 Significant Implications

This research holds significant implications for various stakeholders, including investors, traders, regulators, and policymakers. By exploring the potential of AI, ML, and blockchain technology in decentralized trading markets, this study aims to identify new opportunities for enhancing market efficiency, transparency, and security. The findings of this research can inform the development of innovative trading strategies and tools, as well as contribute to the establishment of appropriate regulatory frameworks that balance innovation with consumer protection and market stability. To address the research questions and test the hypotheses, this study will employ a combination of quantitative research methods. The quantitative approach will involve the collection and analysis of historical market data, as well as the application of various AI and ML techniques, such as logistic regression decision trees, and clustering algorithms. The performance of these models for evaluated using RMSE metrics as the regressor problem. The research will involve a review of relevant literature, as well as gaining insights into the ethical and regulatory issues surrounding the use of AI, ML, and blockchain technology in decentralized trading markets.

3.1.9 Scope and Limitations

While this research aims to make a significant contribution to understanding the potential of AI, ML, and blockchain technology in decentralized trading markets, it acknowledges certain limitations. This study primarily relies on historical market data, which, although abundant, may not completely represent future market trends. Also, while the ML models used may predict certain market trends effectively, they may not account for all potential market movements due to the inherently unpredictable nature of financial markets. Moreover, the views expressed by the market participants and experts in the qualitative research component are subjective and may not encompass the broad spectrum of opinions within the field (Figure 3.1).

3.1.10 Theoretical Framework

The theoretical framework guiding this research is rooted in computer science, AI, ML, and financial economics. From computer science, we adopt the principles of algorithm design and data structures that underpin the AI and ML models used. The principles of pattern recognition and learning systems from AI and ML inform the development and evaluation of these models. From financial economics, we draw on theories about market efficiency, asset pricing, and investor behaviour. This interdisciplinary approach provides a comprehensive framework for understanding the complex interplay between AI, ML, blockchain technology, and decentralized trading markets.

3.1.11 Contribution to the Field

This research aims to add to the growing body of knowledge on the application of AI, ML, and blockchain technology in trading markets. By integrating these technological innovations into trading strategies in decentralized markets, it seeks to offer new insights into market efficiency, transparency, and security. This study's results could inform the design of more sophisticated trading models and tools. It also offers a deeper understanding of the ethical and

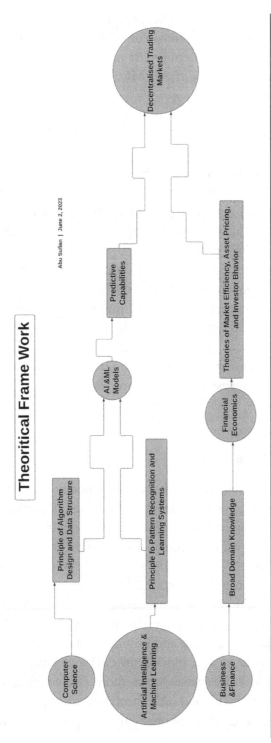

Figure 3.1 Theoretical framework.

regulatory challenges in this arena, potentially guiding policymaking to balance technological innovation with market stability and consumer protection. Furthermore, it might inspire future research in this exciting intersection of AI, ML, blockchain technology, and financial markets.

3.1.12 Outline

The literature review will provide an overview of existing research on AI, ML, and blockchain technology in trading markets, as well as highlight any gaps in knowledge that this study aims to address. The methodology section will describe the research design, data collection, and data analysis procedures in more detail. The results section will present the findings of this study, including the performance of the AI and ML models and any patterns or trends that emerge from the data. The discussion section will interpret these results in the context of the research questions and hypotheses, as well as explore their implications for decentralized trading markets. Finally, the conclusion will summarize the key findings and contributions of this study, as well as suggest areas for future research.

3.2 Literature Review

3.2.1 AI vs Blockchain Technology

Blockchain technology and AI have the potential to work well together and open new applications and business opportunities. The ability to create decentralized autonomous organizations (DAOs) that can function without human intervention is one of the key advantages of fusing blockchain and AI technology. DAOs are businesses that are controlled by smart contracts and can automate a few processes, including decision-making, resource allocation, and conflict resolution. By identifying and reducing threats, AI can also be used to improve the safety and privacy of blockchain systems. AI can also be used to analyse blockchain data and uncover insightful trends that can help businesses make more efficient decisions. The marriage of blockchain and AI, however, also presents several difficulties, including the requirement for specialized knowledge and skills, the potential for new types of centralizations, and the moral ramifications of AI-driven decision-making in DAOs.

The possibility for new types of centralization derives from the fact that the creation of AI models frequently necessitates huge data sets, which may end up being concentrated in the hands of a small number of companies. This data concentration may result in a new type of centralization, whereby companies having access to vast amounts of data hold disproportionately great power and influence inside the AI and blockchain ecosystems. Another issue with AI-driven decision-making in DAOs is its ethical ramifications. The ethics of AI models depend on the data they are trained on, and the AI system may reinforce the biases and prejudices found in the data.

As a result, decisions that are unfair and discriminating might be made, which would be bad for both the individual and society. Academics and decision-makers are striving to create moral guidelines and prescriptions for the creation and use of blockchain and AI technology to allay these worries. Blockchain technology and AI are two of the most inventive and revolutionary technologies of our day. Several industries stand to undergo fundamental change because of these technologies, which also present new economic prospects. People must nevertheless overcome several obstacles to reach their full potential.

This will enable us to harness their transformative power responsibly and ethically. Ultimately leading to innovative solutions that benefit both individuals and society as a whole. Two of the most innovative and impactful technologies in modern times are AI and blockchain technology. They have the power to completely alter the way humans work and live, and they are already making a significant impact on the world [14].

3.2.1.1 How It Works In contrast to AI, which is focused on creating intelligent systems that can observe, learn, and reason like people, blockchain technology aspires to construct decentralized and transparent systems that can securely store and exchange data without the need for middlemen [15]. The use of blockchain and AI technologies together has the potential to promote better levels of innovation and value generation. Using applications like smart contracts, predictive analytics, and fraud detection, among

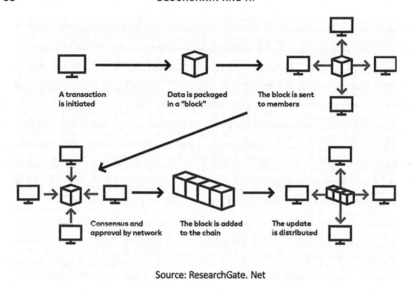

Source: ResearchGate. Net

Figure 3.2 Working process of Blockchain.

[Source: Researchgate.Net]

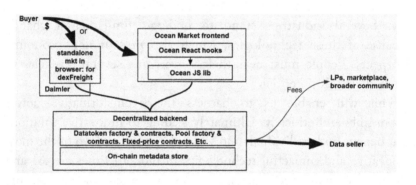

Figure 3.3 A decentralized backend is shared by ocean marketplaces [23].

others, AI may be utilized to increase the effectiveness and precision of blockchain systems [16].

Blockchain technology can offer a safe and centrally controlled framework for AI systems, promoting data privacy, security, and cooperation [17]. Data privacy and security developments are one of the main advantages of merging AI and blockchain technology. Individuals and businesses can have more control over their data and ensure that it is not compromised or misused by adopting blockchain

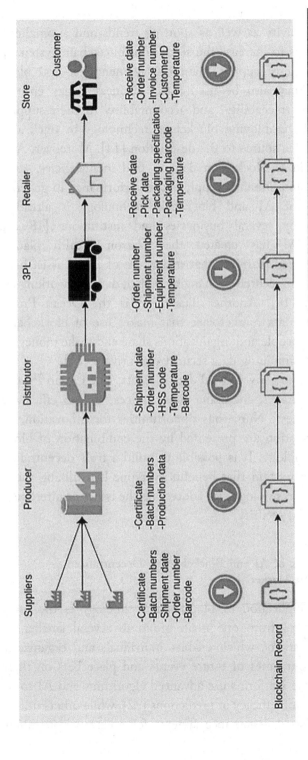

Figure 3.4 Blockchain in supply chain management [21].

technology to store and distribute it [7]. By detecting and stopping fraudulent activity as well as spotting trends and anomalies in the data, AI helps to increase the security of blockchain systems [16].

In the area of supply chain management, AI and blockchain technology may also be used [16]. Businesses may ensure more transparency, traceability, and accountability in their supply chain operations by employing blockchain technology to track and trace things from the source to the destination [14]. Moreover, AI can be used to evaluate blockchain data to find inefficiencies, streamline procedures, and enhance supply chain performance in general.

The use of AI and blockchain technology is already being investigated by several businesses and institutions [18–20]. For instance, IBM has created the Watson Health platform, a blockchain-based platform that makes use of AI to assist healthcare organizations in securely exchanging patient data and working together on research [18]. Another illustration is the Ocean Protocol, a decentralized data marketplace that makes use of blockchain technology to let people and organizations share and make money off their data while maintaining data security and privacy [21].

Compared to conventional data markets, The Ocean Protocol has various advantages, including lower prices, higher efficiency, and greater openness. Numerous opportunities for innovation, growth, and value creation are presented by the combination of blockchain and AI technology. It is possible to build a truly decentralized and intelligent ecosystem that benefits everyone by utilizing the advantages of each technology and addressing the issues of interoperability, scalability, and security.

3.3 Importance of AI and Blockchain in Decentralized Prediction Markets

A hopeful application of blockchain technology is the process of shifting control from one main group to several smaller ones in prediction markets, which enables individuals and organizations to predict the outcomes of future events and place bets on them. For instance, some platforms use advanced algorithms and AI to increase the accuracy and efficacy of predictions [22], while others utiliseutilize blockchain-based governance systems for greater accountability and transparency [23–39]. By utilizing blockchain technology to create a decentralized and transparent platform, prediction markets can provide decision-makers and stakeholders with useful information and insights [10,40]. To ensure the accuracy and dependability of the predictions, prediction markets must also use the capabilities of AI [40].

Decentralized prediction markets gain fundamentally from AI by improving accuracy and efficiency. AI can analyse vast volumes of data, find patterns and trends that humans might miss, and improve forecasting and decision-making. AI can automate a range of prediction market functions, including data collection, processing, and trading, to lessen the risk of fraud and inaccuracy (Figure 3.5).

AI has several advantages for decentralized prediction markets, including increased scalability and interoperability. Traditional prediction markets are often limited in their scope and reach, but AI can help them scale up to meet the demands of large-scale applications and enable interoperability with other platforms and systems. This

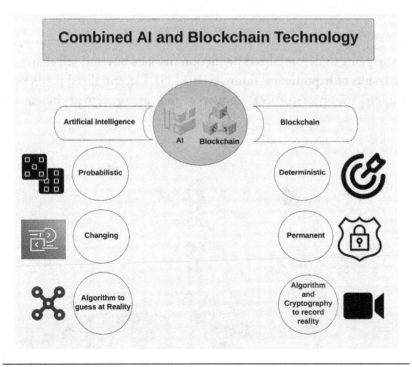

Figure 3.5 Combined AI and blockchain technology [41].

helps prediction markets to become more accessible and efficient while allowing larger volumes of data or users [10,40].

Traditional prediction markets don't typically include AI because they rely on human judgement and expertise [10,40]. The ethical, legal, and regulatory ramifications of integrating AI into prediction markets must be carefully considered. To ensure accountability and transparency, AI-powered prediction markets may give rise to concerns about data privacy, security, and fairness.

Several projects have emerged that investigate the use of AI and blockchain technology in fragmented prediction markets to address these issues [24,39]. AI and blockchain technology has the potential to revolutionize diffused prediction markets by providing greater accuracy, efficiency, scalability, and interoperability. However, integrating AI into prediction markets also requires careful consideration of ethical, legal, and regulatory implications, and could necessitate new frameworks and standards to ensure transparency and accountability (Figure 3.6).

3.4 Fundamentals of Decentralized Market Prediction

3.4.1 What Are the Decentralized Prediction Markets?

Participants in decentralized prediction markets buy and sell shares in the results of hypothetical future events [44]. Decentralized prediction markets use peer-to-peer networks and are governed by smart

Figure 3.6 Both blockchain and AI eliminate the intermediator to enhance transparency [42,43].

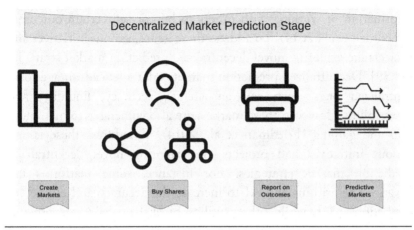

Figure 3.7 Decentralized market prediction stage [35].

contracts in contrast to traditional prediction markets, which are frequently centralized and managed by a single entity [44].

Decentralized prediction markets' transparency and dependability are their main advantages (Friedman et al., 2014). It is impossible for one person or entity to manipulate the market because they run on a blockchain network and all transactions and trades are visible to all participants (Figure 3.7) (Friedman et al., 2014).

Decentralized market predictions leverage the contributions of numerous participants, facilitated by blockchain technology, to generate predictions that are both more precise and transparent in nature. Another benefit of decentralized prediction markets is their capacity to give stakeholders and decision-makers useful data and insights [44]. Prediction markets, which combine the opinions and forecasts of many participants, frequently offer forecasts that are more precise and dependable than those made using more conventional techniques [44]. Decentralized prediction markets are advantageous for accessibility and liquidity (Ethereum Project, 2022). Prediction markets can draw a huge and diversified pool of participants due to its accessibility to anybody with an Internet connection, resulting in increased liquidity and better prices (Ethereum Project, 2022).

Decentralized prediction markets are not without their difficulties, such as vulnerability to manipulation and gaming (Friedman et al., 2014). People may try to manipulate the market by disseminating false information or working with others to buy or sell certain outcomes (Friedman et al., 2014). To address these issues, various initiatives and projects are exploring novel decentralized prediction market strategies [24,39]. Decentralized prediction markets offer a few advantages over centralized ones, such as transparency, accuracy, liquidity, and accessibility. However, these markets are also vulnerable to manipulation and gaming (Friedman et al., 2014). To address these issues, various initiatives and projects are exploring novel decentralized prediction market strategies. For instance, some platforms use advanced algorithms and AI to increase the accuracy and efficacy of predictions [22], while others utilize blockchain-based governance systems for greater accountability and transparency [24,39].

In addition to these strategies, several tools and resources are available to assist participants in navigating decentralized prediction markets. For instance, some platforms provide tools for data visualization and user-friendly interfaces to aid users in making wise decisions (Ethereum Project, 2022). Others offer educational materials and discussion boards to help users understand the market better or connect with other participants. Furthermore, graphs/charts displaying the volume/price of different predictions now (Ethereum Project, 2022) or tables outlining the most popular predictions with their present costs/ results can be used as explanatory aids [24,39]. Decentralized prediction markets have a lot of potential that could be harnessed by utilizing various strategies such as those mentioned above.

3.5 Advantages of Decentralized Prediction Markets over Traditional Prediction Markets

3.5.1 Here Are Some of the Key Advantages

Decentralized prediction markets offer several key essentials over traditional prediction markets, including transparency, accuracy, liquidity, accessibility, lower costs, decentralized governance, and opportunities for innovation. Transparency is ensured using a decentralized blockchain network which makes all transactions visible to all participants. This also makes it impossible for any one individual or entity to manipulate the market. Additionally, these markets can often provide more accurate and reliable forecasts than traditional methods by aggregating the opinions and predictions of many participants.

Table 3.1 Decentralized vs Traditional Prediction Markets [40,45,46]

ASPECT	DECENTRALIZED PREDICTION MARKETS	TRADITIONAL PREDICTION MARKETS
Openness	High (open to everyone)	Low (restricted access)
Transparency	High (visible on blockchain)	Low (less transparent)
Immutability	High (immutable records)	Low (modifiable records)
Anonymity	High (anonymous trading)	Low (requires identification)
Access	24-Jul	Limited business hours

Furthermore, they are open to anyone with an Internet connection and often have lower transaction costs due to their decentralized nature. Last but not least, some decentralized prediction markets are governed by smart contracts and decentralized governance systems which ensure transparency and accountability (Table 3.1).

3.6 How Does Blockchain Technology Enable Decentralized Prediction Markets?

3.6.1 Decentralization

Blockchain technology, in accordance with Hanson [20], promotes the development of decentralized prediction markets by offering a safe and open trading and forecasting platform. Decentralization guarantees that no one organization has total authority over the market, strengthening its defences against manipulation. Blockchain technology's immutability is a crucial component that lends decentralized prediction markets transparency and dependability. Data that has been stored on the blockchain cannot be changed or removed, according to [24,39].

3.6.2 Smart Contracts

Smart contracts are an important feature of blockchain technology that enables the creation of automated market makers and other governance systems. According to Hanson [20], smart contracts are self-executing contracts with the terms of the agreement written into the code (Figure 3.8).

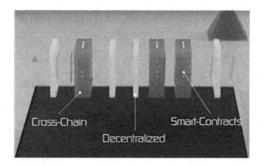

Figure 3.8 The stage of smart-contractors [39].

3.6.3 Tokenization

Blockchain technology enables the use of tokens as a medium of exchange in prediction markets. According to Augur [24,39], tokens can represent a prediction of the outcome of an event or can be used as collateral to create synthetic assets. It has a simple definition; Tokenization represents a process which creates digital tokens using a smart contract on a blockchain. These tokens are utilized to reveal the ownership of various assets, equities, finances, and resources related to a project (Figure 3.9).

3.6.4 Security

The security of blockchain technology is essential for the success of decentralized prediction markets. As noted by Hanson [20],

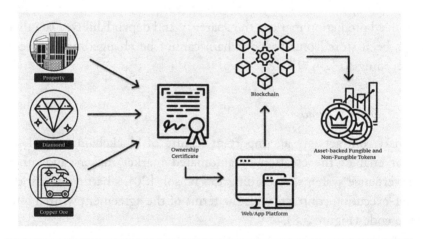

Figure 3.9 Tokenization [47].

blockchain technology uses cryptographic techniques to secure transactions and prevent fraud.

3.7 ML in Trading Strategies

3.7.1 Basics of ML for Trading Strategies

Because it can analyse vast amounts of data and find patterns, ML has gained popularity in trading strategies. The fundamentals of ML for trading strategies, along with the pertinent mathematical terms, are as follows:

Supervised learning is a type of ML in which the algorithm picks up knowledge from labelled data. This implies that the algorithm is trained using historical data with well-known results in the context of trading strategies. Decision trees, logistic regression, and linear regression are common supervised learning methods used in trading strategies.

Unsupervised learning is a type of ML in which the algorithm picks up knowledge from unlabelled data. This means that in the context of trading strategies, the algorithm is used to find patterns in data without knowing the results in advance. Principal component analysis and clustering are two common unsupervised learning methods used in trading strategies.

Reinforcement learning is a type of ML in which the algorithm picks up knowledge by making mistakes. This implies that the algorithm is trained to maximize a reward function based on its actions in the context of trading strategies. Q-learning and deep reinforcement learning are two common reinforcement learning methods used in trading strategies.

A common issue in ML is **overfitting**, where the algorithm fits the training data too closely and is unable to generalize to new data. Poor performance in real-world situations may result from this. To avoid overfitting in trading strategies, regularization techniques like L1 and L2 regularization are frequently used.

Cross-validation is a technique used to assess how well a ML algorithm performs on fresh data. To test an algorithm's generalizability to new data, it is tested on a different dataset in the context of trading strategies. The K-fold cross-validation and leave-one-out cross-validation are two frequently used cross-validation methods in trading strategies.

A **potent set of tools** is available for creating trading strategies through ML. Trading professionals can use these techniques to extract insights from massive amounts of data and create more successful trading strategies by understanding the fundamentals of ML.

3.8 Techniques for Analysing Data in Prediction Markets

People are using market conditions forecasts more frequently to create predictions about uncertain events. To analyse data in prediction markets and make reliable projections, mathematical methods must be used. Traders and analysts can benefit from a variety of mathematical tools, such as probability theory, Bayesian inference, regression analysis, ML, and Monte Carlo simulation, to gather critical data and create more accurate forecasting tactics. Nowadays, a variety of mathematical techniques are used by traders.

Probability theory: According to Bernardo and Smith [27], probability theory is a key idea in prediction markets. Probability is best defined as the likelihood that an event will occur and can be expressed as a number between 0 and 1. In prediction markets, probabilities are routinely employed to depict the likelihood that a particular outcome will occur (Figure 3.10).

Bayesian inference: According to Bernardo and Smith [27], it is a statistical method for revising beliefs considering new information.

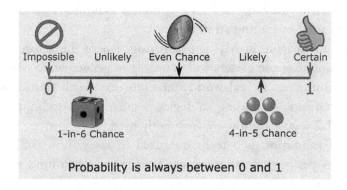

Figure 3.10 Probability [39].

Bayesian inference can be used in prediction markets to revise the probability of the hypothesis of new data. In mathematical terms, it can be represented as:

$$P(H|E) = [P(E|H) * P(H)]/P(E)$$

where $P(H|E)$ is the posterior probability of the hypothesis H given evidence E, $P(H)$ is the prior probability of the hypothesis, $P(E|H)$ is the likelihood of the evidence given the hypothesis, and $P(E)$ is the marginal likelihood of the evidence.

Regression analysis: According to Hastie et al. [25], regression analysis is a statistical technique used to examine the relationship between two or more variables. Regression analysis can be used in prediction markets to find variables that are related to an event's outcome. The basic linear regression model can be expressed mathematically as:

$$y = mx + c + \varepsilon$$

where y is the dependent variable, x is the independent variable, c is the intercept, m is the slope, and ε is the error term. The slope coefficient (m) represents the change in the dependent variable (y) for a unit change in the independent variable (x). The intercept (c) represents the value of y when x is zero. The error term (ε) represents the variability in y that is not explained by the independent variable x. This is also known as the method of least squares (Figure 3.11).

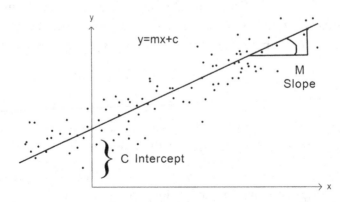

Figure 3.11 Regression analysis.

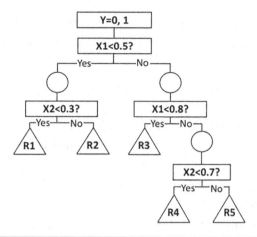

Figure 3.12   Sample decision tree based on binary target variable Y [48].

Decision tree: Data in prediction markets can be analysed using ML techniques like decision trees, support vector machines, and neural networks [25]. These methods can be used to find patterns in data and forecast outcomes in the future (Figure 3.12).

Monte Carlo simulation is a statistical method for simulating a wide range of potential outcomes for a specific event [28]. Monte Carlo simulation can be used in prediction markets to calculate the likelihood of various outcomes based on a variety of potential outcomes. The mathematical model for Monte Carlo simulation can be expressed as follows:

Let X be a vector of input variables, and Y be a vector of output variables. Let f(x,y) be a model that describes the relationship between X and Y. Then, the Monte Carlo simulation involves generating n random samples of X from a probability distribution, and using these samples to simulate the behaviour of the system:

$$Y_i = f(X_i, \quad i = 1, 2 \ldots \ldots \ldots \ldots \ldots n)$$

The results of the simulation can then be analysed to estimate the behaviour of the system under different conditions, such as changes in the input variables or changes in the model itself.

3.9 Advantages of Using ML in Trading Strategies

Some advantages of using ML in trading strategies are as follows:

Improved accuracy: Large volumes of data may be accurately analysed by ML algorithms, which can result in more precise predictions and better trading choices [49]. This is crucial when it comes to high-frequency trading because accurate and timely judgements are required.

The mean squared error (MSE) loss function is a mathematical formula that is frequently used in ML for trading strategies and is used to calculate the difference between projected and actual prices.

The formula for the MSE is:

$$MSE = (1/N) * \sum (P_t + 1 - P_{t\_pred})^2$$

where

P_t+1 is the actual price of the security at the next time step.
$P_{t\_}pred$ is the predicted price of the security at the next time step
N is the total number of predictions.

The goal of the ML algorithm is to minimize the MSE loss function during training, by adjusting the model's parameters to make more accurate predictions of future prices.

Let us consider a dataset of historical market prices for a particular security or financial instrument, denoted by P. Each element in this dataset, represented by P_i, reflects the price of the security at a specific moment in time. A trading strategy's objective is to forecast the security's future price, P_{t+1}, using previous data.

The neural network takes the historical price data as input and outputs a predicted price for the next time step, P_{t+1}.

Reduced bias: reducing bias is important in developing accurate and reliable trading strategies using ML. ML algorithms are not influenced by human biases, which can lead to more objective and consistent trading decisions [50]. This can be particularly important in the context of financial markets, where emotions and biases can lead to suboptimal decisions. In trading, bias can lead to inaccurate predictions and flawed trading decisions, which can result in losses. If I apply mathematical formulas, then precise and informed decisions can be made.

The models computed root-mean-squared errors fit the dataset displayed in Figure 3.13. When the models get more complicated,

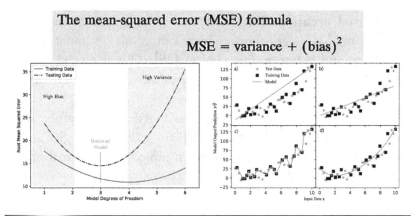

Figure 3.13 Root mean square error vs model degrees of freedom [50].

the training data keep getting smaller, which is an example of overfitting. Yet, the RMSE considerably rises with increasingly sophisticated models when the model is tested against a test dataset.

There are several ways that reducing bias can help in trading strategies:

- Improving Predictive Accuracy
- Increasing Robustness
- Enhancing Generalization (e.g., k-fold cross-validation)

Improved efficiency: ML algorithms can analyse large amounts of data quickly and efficiently, which can save time and resources [49]. This can be particularly important in the context of high-frequency trading, where decisions must be made quickly to take advantage of market opportunities. One commonly used metric for comparing trading strategies is the Sharpe ratio.

The Sharpe ratio measures the excess return generated by a portfolio relative to the risk-free rate per unit of portfolio risk. The formula for the Sharpe ratio is:

$$\text{Sharpe ratio} = (Rp - Rf)/\sigma p$$

where

Rp is the expected return of the portfolio.
Rf is the risk-free rate of return.
σp is the standard deviation of the portfolio's excess returns.

In the context of ML-based trading strategies, the Sharpe ratio for a portfolio can be calculated that is generated using predictions from an ML model and compared to the Sharpe ratio of a portfolio generated using a traditional trading strategy.

If the Sharpe ratio of the ML-based portfolio is higher than that of the traditional portfolio, then it can be said that the ML-based trading strategy is more efficient in generating returns. Therefore, the formula for demonstrating the improved efficiency of a ML-based trading strategy is:

$$\text{Sharpe ratio (ML)} > \text{Sharpe ratio (traditional)}$$

Sharpe ratio (ML) is the Sharpe ratio of the portfolio generated using an ML-based trading strategy. Sharpe ratio (traditional) is the Sharpe ratio of the portfolio generated using a traditional trading strategy.

Ability to analyse complex data: ML algorithms can analyse complex data sets, such as text and images, which can provide valuable insights for trading decisions (Scherer, Müller, & Behnke, 2010). For example, natural language processing techniques can be used to analyse news articles and social media posts to identify trends. and sentiments related to specific companies or industries.

Here is an example of how ML algorithms can analyse complex data in trading strategies using a mathematical model:

Suppose we have a dataset of historical price data, news articles, and social media sentiment for a particular stock. We want to use this data to predict the future price movement of the stock. We can use a machine learning algorithm, such as a neural network, to analyse the data and make a prediction.

Let's denote the historical price data as P, the news articles as N, and the social media sentiment as S. We can represent this data as a matrix X:

$$X = [P, \quad N, \quad S]$$

We also have the corresponding target variable, the future price movement of the stock, which we denote as Y.

We can use the ML algorithm to learn a function f(X) that maps the input data X to the output Y. This function f(X) can consider the complex relationships between the different types of data, such as the effect of news articles and social media sentiment on the stock price. Once we have trained the model, we can use it to make predictions on new data. For example, if we have new news articles and social media sentiment data, we can use the model to predict the future price movement of the stock.

Adaptability: ML algorithms can modify trading tactics in response to shifting market conditions [49]. This can be especially crucial in the setting of financial markets that are constantly changing and unpredictable.

Adaptability in trading techniques based on ML refers to the model's capacity to adapt to changing market conditions and learn from new data. One mathematical model that demonstrates adaptability is the online learning algorithm. The stochastic gradient descent algorithm is an illustration of an online learning tool (SGD). The formula for the update rule for SGD is:

$$\theta t + 1 = \theta_t - \alpha * \nabla L(y_t, \ f(x_t; \ \theta^t))$$

where

θ_t is the model's parameter at time step t.

α is the learning rate, which determines the step size of the update.

L is the loss function, which measures the difference between the predicted and actual values.

y^t is the actual value at time step t.

$f(x_t; \theta t)$ is the predicted value at time step t, based on the current parameters θt.

$\nabla L(yt, f(x_t; \theta_t))$ is the gradient of the loss function with respect to the parameters θt.

This update rule allows the model to continuously adjust its parameters based on new data, allowing it to adapt to changing market conditions. Trading professionals and analysts can get insightful information and create more successful trading strategies by making use of these benefits.

3.10 Developing Trading Strategies for Decentralized Prediction Markets

To demonstrate the entire project, the following process had to come up with Python. In Python, implementation has been done as the entire project to develop trading strategies for Dispersed market prediction (Figure 3.14).

3.10.1 Conceptual Framework

3.10.2 Data Collection and Pre-Processing

In this section, the dataset has been used from Yahoo Finance and Kaggle to get a glimpse of real-life analysis using ML, AI and blockchain tools by Python code. This dataset is a secondary source that is collected from websites. The data presented are from secondary data but not from web scraping (Figure 3.15).

The missing and duplicate values in the datasets were dropped while the continuous variables were scaled using standardization to ensure the same features.

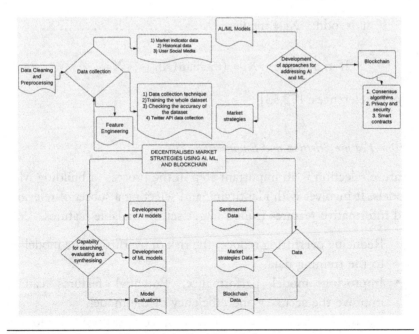

Figure 3.14 Conceptual framework for the data analysis.

Data Preprocessing

```
#Handling NAs
df.dropna(inplace=True)
```

Figure 3.15 Data processing.

The dropna () function is the most general method, which is used for handling missing data, and the inplace=True parameter ensures that the changes are made to the original dataset instead of creating a copy. The dataset can be used for analysis and modelling more effectively by deleting the rows with missing data because missing values can interfere with some algorithms and analyses.

Handling missing data is to impute the missing values with a measure of central tendencies, such as the mean or median of the non-missing values in the same column. This can be expressed mathematically as:

For mean imputation:

$$X_i = (X_1 + X_2 + \ldots + X_n)/n$$

For median imputation:

$$\text{If } n \text{ is odd: } Xi = \text{median } (X_1, \ X_2, \ldots\ldots\ldots\ldots\ldots\ldots X_n)$$

$$\text{If } n \text{ is even: } X_i = (\text{median}(X_{(n/2)}, \ X_{(n/2+1)})$$

Model references: [52,53]

3.10.3 Feature Selection and Engineering

Feature selection is an important step in the process of building ML models. It involves with identifying and selecting a subset of relevant and informative features from a larger set of available features.

- Reducing overfitting reduces the risk of overfitting the model to the training data.
- Improving model performance: Potential features can improve the accuracy and efficiency of the model.
- Saving computation time and resources: Feature selection can assist in lowering the computational cost and length of

time needed to train the model by minimizing the number of dimensions in the dataset.

- Enhancing interpretability: by reducing superfluous or redundant features.

To perform feature engineering, exploratory data analysis should be done to comprehend the distribution and interaction of the features, identify potential sources of bias or noise, and choose the best feature engineering strategies to solve these problems.

3.10.4 Market Indicators

Market volatility: Daily, weekly, or monthly volatility measures such as standard deviation of returns, average true range, or Bollinger bands.

Liquidity: Measures of trading activity such as trading volume, bid ask spread, or open interest.

Trading volume: Total volume, volume change, or volume ratio measures.

There are several approaches for feature engineering including correlation matrices and PCA. Principal component analysis is used to remove dimensionality from a dataset. By figuring out the standard deviation of the daily returns, I can build a variable for the daily volatility. Similarly, I may develop variables for trade volume and liquidity using the metrics.

Below feature engineering methods and how they can be deployed in Python has been shown. This is just a general way of coding in Python implementation for the decentralized trading market prediction task.

3.10.5 Correlation

The matrix shows how each unique pair of values in a table correlates with one another [50]. It is a potent tool for finding and displaying trends in the provided data as well as for summarizing a sizable dataset. A correlation matrix defined as a heatmap can be used as a feature selection method to identify strongly associated variables and remove redundant features. The correlation only measures linear

```
import pandas as pd
import numpy as np
# Load the dataset
df = pd.read_xlsx('Market_Indicator_Data_combined.xlsx')

# Convert the date variable to datetime
df['date'] = pd.to_datetime(df['date'])

# Create variables for market volatility
df['daily_volatility'] = df['daily_returns'].rolling(window=30).std()
df['weekly_volatility'] = df['weekly_returns'].rolling(window=4).std()
df['monthly_volatility'] = df['monthly_returns'].rolling(window=12).std()

# Create variables for liquidity
df['bid_ask_spread'] = df['ask_price'] - df['bid_price']
df['open_interest_ratio'] = df['open_interest'] / df['trading_volume']

# Create variables for trading volume
df['volume_change'] = df['trading_volume'].pct_change()
df['volume_ratio'] = df['trading_volume'] / df['open_interest']

# Select the most important features using feature selection methods
X = df.drop(['date', 'prediction'], axis=1) # Remove non-feature variables
y = df['prediction']
```

Figure 3.16   Feature engineering process in Python, Jupiter Notebook.

relationships between variables, so it does not account for non-linear relationships between variables. Currently, most researchers choose it for exploratory analysis rather than feature selection (Figures 3.17 and 3.18) [50].

Correlation analysis within the heatmap has been conducted to assess the relationships between market indicators, social media sentiment, economic indicators, and user behaviour in decentralized prediction markets.

3.10.5.1 Trading_Volume and Trading_Volume   The correlation of a variable with itself is always 1. Therefore, the diagonal of the matrix from the top left to the bottom right is filled with 1 s.

```
#Correlation plot
sns.heatmap(df.corr())
```

Figure 3.17   Code of heat map as seen in Python.

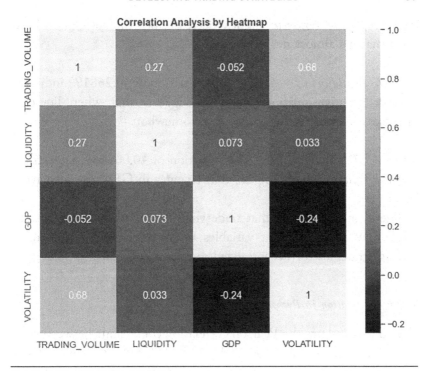

Figure 3.18 Correlation matrix.

3.10.5.2 Trading_Volume and Liquidity A correlation of 0.112113 suggests a weak positive relationship. This means that when the trading volume tends to increase, liquidity tends to slightly increase as well, although this correlation is weak.

3.10.5.3 Trading_Volume and GDP A correlation of 0.111590 also indicates a weak positive relationship. This suggests that when the trading volume increases, the GDP might slightly increase as well, but again, this correlation is weak.

3.10.5.4 Trading_Volume and Volatility A correlation of 0.500249 suggests a moderate positive relationship. This implies that when the trading volume increases, the volatility tends to increase to a certain extent as well.

3.10.5.5 Liquidity and GDP A correlation of –0.031011 indicates a very weak negative relationship. This suggests that when liquidity increases, GDP tends to slightly decrease, but the correlation is so weak that it's almost negligible.

3.10.5.6 Liquidity and Volatility   A correlation of −0.268192 indicates a weak negative relationship. This suggests that when liquidity increases, volatility tends to decrease somewhat.

3.10.5.7 GDP and Volatility   A correlation of −0.004145 suggests no linear correlation. This implies that changes in GDP do not have a linear effect on volatility.

These findings imply that trade volume, volatility, liquidity, and GDP may all be relevant variables to consider when formulating trading strategies for decentralized market predictions.

3.10.6 Calculation in Python

The code calculates the Pearson correlation coefficients between each pair of numerical variables in the provided dataset. The Pearson correlation is a statistical measure of the linear relationship between two variables that are calculated in Python. Its values range from −1 (perfectly negative correlation) to 1 (perfectly positive correlation), with 0 denoting no correlation. This code can be applied to a variety of tasks, including discovering highly correlated variables, investigating the relationships between variables, and choosing a subset of variables for modelling or analysis (Table 3.2).

3.10.7 Exploratory Analysis

3.10.7.1 Volatility vs Liquidity   Volatility refers to the degree of fluctuation in an asset's price over time. An asset with high volatility

Table 3.2 Python Code: Pearsoncorr=df.corr(method='pearson') pearsoncorre

	TRADING VOLUME	LIQUIDITY	GDP	VOLATILITY
Trading volume	1	0.23388	−0.046511	0.673695
Liquidity	0.23388	1	0.068832	0.033218
GDP	−0.046511	0.068832	1	−0.238131
Volatility	0.673695	0.033218	−0.238131	1

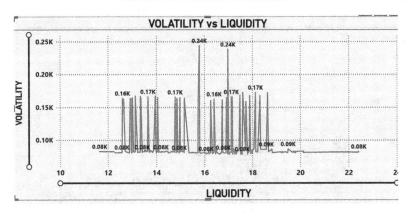

Figure 3.19 Volatility vs liquidity.

experiences significant price changes, often in short periods, while an asset with low volatility tends to have a more stable price. Volatility can be caused by various factors such as changes in supply and demand, economic and political events, or even human emotions like fear and greed (Figure 3.19).

Liquidity, on the other hand, refers to the ease and speed with which an asset can be bought or sold in the market without significantly affecting its price. An asset with high liquidity has many buyers and sellers, so transactions can be executed quickly and at a fair price. In contrast, an asset with low liquidity may have a limited number of buyers and sellers, making it difficult to trade or resulting in transactions executed at unfavourable prices.

3.10.7.2 Trading Volume vs Volatility The total number of shares or contracts exchanged in a specific market during a specific time is referred to as "trading volume". Larger trade volumes typically signify more market activity and interest in the item, which might help to explain why volatility is higher (Figure 3.20).

In other words, the price of an item may change more as more individuals buy and sell it. Assets with low volatility typically have more stable prices, whereas assets with high volatility frequently undergo major price swings in short periods (Figure 3.21).

3.10.7.3 GDP vs Volatility A strong GDP may indicate a good economy, which could lead to increased investment and higher

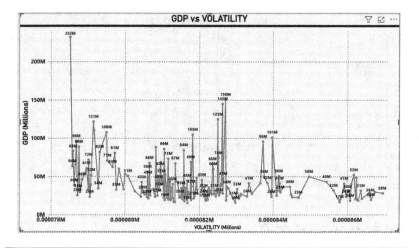

Figure 3.20 Trading volume vs volatility.

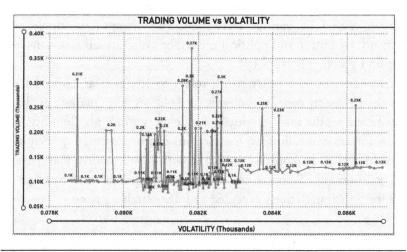

Figure 3.21 GDP vs volatility.

demand for assets, potentially driving prices up and reducing volatility. Decentralized trading markets are often subject to a wide range of other factors that can influence volatility, such as supply and demand changes, regulations or policies, news events, and technological developments. These factors may be more directly relevant to the behaviour of the decentralized trading market than changes in GDP (Table 3.3–3.5).

Table 3.3 Summary Table for Trading Volume Python Code: df['Trading Volume']. describe()

	TRADING VOLUME
count	235
mean	106.010976
std	13.634665
min	80.618034
25%	96.916748
50%	102.36145
75%	117.761375
max	133.096725
Name	Trading Volume
dtype	Float64

Table 3.4 Python Code: df['LIQUIDITY']. describe ()

count	235
mean	15.84881
std	2.186753
min	11.65
25%	13.96
50%	15.98
75%	17.48
max	22.41
Name	LIQUIDITY
dtype	Float64

Table 3.5 Python Code: df ['GDP']. describe ()

	GDP
count	235
mean	34469710
std	22410820
min	12007600
25%	23566850
50%	28688600
75%	36662500
max	2.32E+08
Name	GDP
dtype	Float64

3.10.8 Descriptive Statistical Summary

The distribution plot shows the change in the data distribution as shown in Figure 3.16. Secondly, the Seaborn distplot shows the total distribution of continuous data variables. In Python, the distplot with several modifications is shown using the Seaborn module. The Distplot represents the data by combining a line with a histogram. Histogram 1 shows the distribution plot for the variable "trading volume". Figures 3.2, 3.3 and 3.4 shows the distribution plot for Liquidity, GDP, and Volatility respectively.

A descriptive statistics table was presented to summarize four variables: trading volume, liquidity, GDP, and volatility. Figure 3.23 shows that there were 364 observations for trading volume, 251 observations for liquidity, 251 observations for GDP, and 241 observations for volatility. The mean trading volume was 98.37, the mean liquidity was 15.79, the mean GDP was 34,575,040, and the mean volatility was 82.00. The standard deviation of trading volume was 16.23, the standard deviation of liquidity was 2.18, the standard deviation of GDP was 21,912,470, and the standard deviation of volatility was 1.98. The minimum and maximum values of trading volume were 74.31 and 133.10, respectively. The minimum and maximum values of liquidity were 11.65 and 22.41, respectively. The minimum and maximum values of GDP were 12,007,600 and 232,316,600, respectively. The minimum and maximum values of volatility were 78.52 and 86.96, respectively. The first quartile trading volume was 83.69, and the third quartile was 108.04 (Figure 3.25).

3.10.9 Types of ML

3.10.10 Splitting the Dataset

```
# Split the data into training, validation, and testing sets
X = df_poly. drop('TRADING_VOLUME', axis=1)
y = df_poly['TRADING_VOLUME']
X_train, X_test, y_train, y_test = train_test_split (X, y, test_size=0.3,
random_state=42)
X_train, X_val, y_train, y_val = train_test_split (X_train, y_train,
test_size=0.2, random_state=42)
```

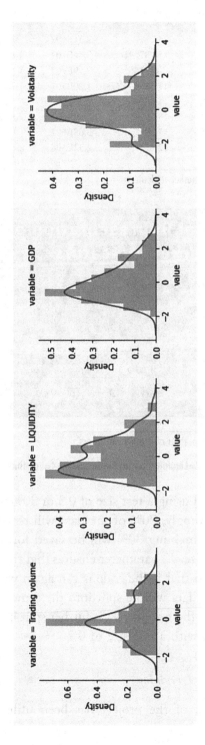

Figure 3.22 Distribution of the variables after scaling.

Statistical Summary	Trading volume	Liquidity	GDP	Volatility
Count	364	251	2.51E+02	241
Mean	98.366245	15.787291	3.46E+07	82.002988
Std	16.23278	2.176146	2.19E+07	1.984542
Min	74.312927	11.65	1.20E+07	78.519997
25%	83.686569	13.865	2.37E+07	80.830002
50%	98.556893	15.79	2.90E+07	81.699997
75%	108.041652	17.39	3.68E+07	82.650002
Max	133.096725	22.41	2.32E+08	86.959999

Figure 3.23 Summary of the market indicators.

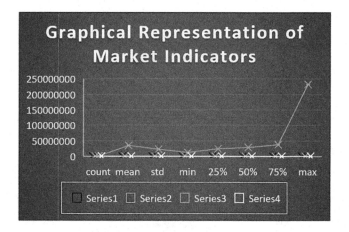

Figure 3.24 Graphical representation of market indicators.

Series1=Trading Volume, Series2=Liquidity, Series3=GDP, Series4=Volatility.

The dataset is divided using a test size of 0.3 or 30% into training and testing sets. Accordingly, 70% of the data will be used to train the model, while the remaining 30% will be saved for performance testing. The random_state=42 parameter ensures that the data is split in a reproducible way so that if the code is run again with the same random state, the same data will be split into the same training and testing sets. Afterwards, the training set is further split into a training set and a validation set with a test size of 0.2.

3.10.11 Model Selection and Training

The problem statement of the project has been utilized for ML model selection,

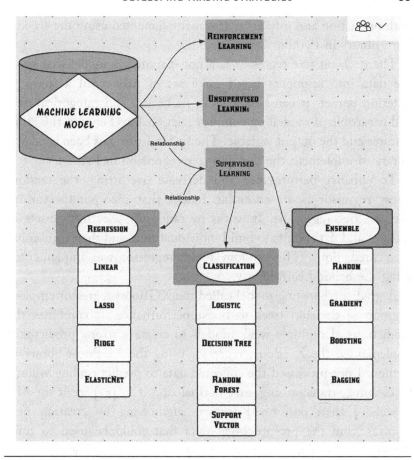

Figure 3.25 Types of machine learning models.

Since market indicators were present in the dataset and could be properly analysed, I decided to apply a regressor model to forecast trading volume. This kind of model is appropriate for trend analysis and can be used to analyse historical data on market indicators. It can be used to find trends and connections that will affect future trading volume directly. Stakeholders can use this information to make decisions based on anticipated future changes in trading volume. So that's why I have chosen the four regression models to get a prediction on trading volume. The models include linear regression, decision tree regressor, random forest regressor, and XGBoost regressor.

In addition, the linear regression model is a simple, yet powerful algorithm that fits a linear equation to the data and can be used for both prediction and inference. It was implemented using the sci-kit-learn library in Python.

The decision tree regressor is a non-parametric model that splits the data into segments based on a set of rules until a stopping criterion is met. It can handle both numerical and categorical data and is capable of modelling complex relationships between the input features and the output variable. The Scikit-Learn has been used as a library to implement the decision tree regression in Python.

To enhance performance and decrease overfitting, the random forest regressor is an ensemble model that incorporates various decision tree regressors. It works by randomly selecting subsets of features and data points to build individual trees and then combining their predictions. The random forest regressor was implemented using the Scikit-Learn library.

A gradient-boosting model called the XGBoost regressor employs a group of decision trees to boost performance. It combines the predictions of multiple weak models to create a strong predictor. I implemented the XGBoost regressor using the XGBoost library in Python. I pre-processed the collected data to predict trading volume by cleaning, transforming, and normalizing it to prepare it for ML models. I then performed feature engineering by creating new features from the pre-processed data that could be used to train the models. I evaluated the performance of each model using the R-squared score and root-mean-squared error metrics.

3.10.12 Model Selection in Mathematical Concept

Using statistical measures like the Akaike Information Criterion (AIC) or the Bayesian Information Criterion (BIC) is a typical approach, claims the mathematical model [54]. These metrics assess how well a model fits the dataset while penalizing models with higher levels of complexity. For example, the AIC is calculated as $AIC = 2k - 2ln(L)$, where k is the number of parameters in the model and L is the likelihood of the data given in the model. The best fit for the data is generally regarded as being the model with the lowest AIC value.

On the other hand, using cross-validation techniques, the data is split into training and testing sets, and different models are trained and evaluated on the testing set. The most suitable model is that which has the lowest prediction error on the testing part [55,56].

3.10.13 AI and ML Models

According to [65], trading volume is an important indicator of market liquidity, and it influences investors' decisions regarding buying or selling shares. In this study, I have explored using ML models to predict trading volume and using various economic indicators, such as GDP, liquidity, and volatility.

To predict the financial trading volume, four regression models, namely linear regression, decision tree regressor, random forest regressor, and XGBoost regressor, will be implemented in predicting the trading volume. I got below different models' results (Table 3.6).

Four regression models were fit to predict the trading volume of a financial instrument based on the features of liquidity, GDP, and volatility. The linear regression model achieved a perfect R-squared score of 1.0 and a very low RMSE of 2.331e–13. The decision tree regressor achieved an R-squared score of 0.991 and an RMSE of 0.090. The random forest regressor achieved an R-squared score of 0.998 and an RMSE of 0.041. Finally, the XGBoost regressor achieved the highest R-squared score of 0.999 and the lowest RMSE of 0.033. Therefore, the XGBoost regressor is recommended for predicting the trading volume of the financial instrument based on the features of liquidity, GDP, and volatility.

Table 3.6 Performance Evaluation and Optimization

DATASET		SIZE
Training set size		107
Validation set size		27
Testing set size		58
MODEL	R^2 SCORE	RMSE
Linear Regression	1	3.75E-13
Decision Tree Regressor	0.992585	0.080396
Random Forest Regressor	0.995396	0.06335
XGBoost Regressor	0.996261	0.057091

3.10.13.1 Multi-Faceted View Analysis of Each Model's Strengths and Weaknesses All Model Result in Scatter Plot within Bivariate Analysis (Figure 3.26).

3.10.13.2 Result in Histogram To identify the model's error distribution and systematic bias in the model's predictions (Figure 3.27).

3.10.13.3 Model Result in Line Plot To identify potential issues with the model, such as non-linearity, heteroscedasticity, or outliers (Figure 3.28).

3.10.14 Sentiment Analysis

A secondary dataset has been used from Kaggle for this sentiment analysis. The dataset includes tweets from the top 25 most popular stock tickers on Yahoo Finance between 30 September 2021, and 30 September 2022. The analysis of Stock Tweets for Sentiment Analysis and Prediction aims to determine market trends and investor sentiment by examining the language and emotions expressed in these tweets.

3.10.14.1 Dataset Description

Date – date and time of the tweet

Tweet – the full text of the tweet

Stock Name – full stock ticker name for which the tweet was scraped.

Company Name – full company name for the corresponding tweet and stock ticker

In the context of decentralized trading market prediction, sentiment analysis is a form of AI that involves using algorithms and ML techniques to analyse and classify the emotional tone of text data such as social media posts, news articles, or online reviews. It can be used to determine how traders and investors feel generally about a specific cryptocurrency or blockchain project. For example, if many traders and investors are bullish or optimistic about a particular cryptocurrency or blockchain project, then its price may increase due to increased demand. Conversely, if sentiment is negative, the price may decrease.

For this sentiment analysis by using the AI model, I train a logistic regression model on the training set. I got an overall accuracy and

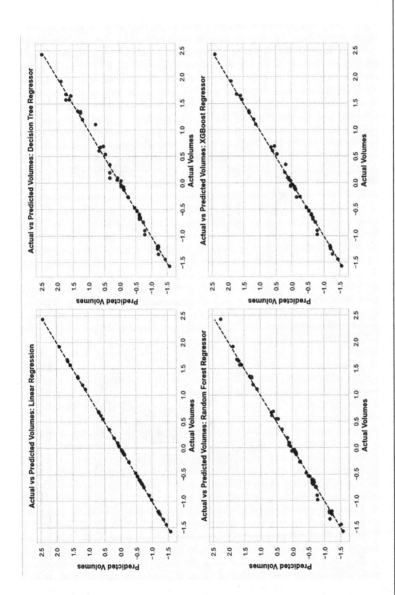

Figure 3.26 Bivariate analysis.

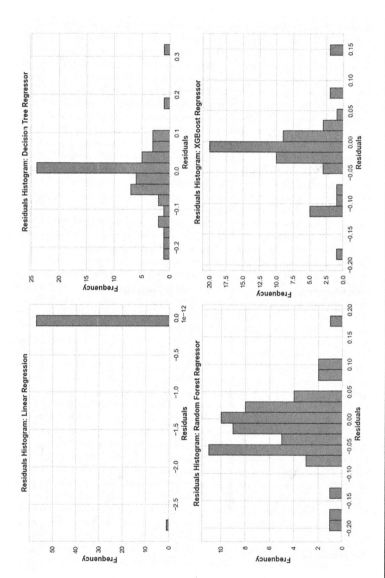

Figure 3.27 Model result in histogram analysis.

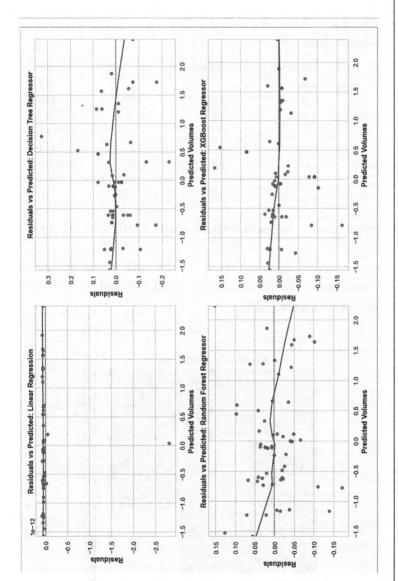

Figure 3.28 Heteroscedasticity analysis.

Table 3.7   Accuracy and Classification Report of an AI Model

Python Code:
Y_pred=lr. predict(x_test)
Acc=accuracy_score (y_test, tes_pred)
Cr= classification_report (y_test, y_pred)
Print ("accuracy, acc)
Print ("classification report:\n", cr)

CLASS	PRECISION	RECALL	F1-SCORE	SUPPORT
negative	0.93	0.87	0.9	2691
neutral	0.95	0.98	0.96	6052
positive	0.97	0.96	0.96	7416
accuracy			0.95	16159
macro avg	0.95	0.94	0.94	16159
weighted avg	0.95	0.95	0.95	16159

classification report by performance metrics on this AI model (Table 3.7).

Above the overall accuracy result is 95% and model evaluation performance metrics like precision 95%, recall 95% and f-1 score 95% as weighted avg (Figure 3.29).

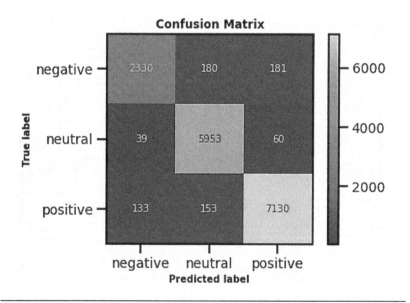

Figure 3.29   Model result evaluation performance by confusion metrics.

- **Negative**: Out of all the samples which were actually Negative, 2,330 were correctly predicted as Negative (True Negative), 180 were incorrectly predicted as Neutral, and 181 were incorrectly predicted as Positive.
- **Neutral**: Out of all the samples which were actually Neutral, 5,953 were correctly predicted as Neutral (True Neutral), 39 were incorrectly predicted as Negative, and 60 were incorrectly predicted as Positive.
- **Positive**: Out of all the samples which were actually Positive, 7,130 were correctly predicted as Positive (True Positive), 133 were incorrectly predicted as Negative, and 153 were incorrectly predicted as Neutral.

3.10.14.2 Result Visualization   Visualization of the distribution of the model's confidence levels (Figures 3.30 and 3.31).

1. **High confidence**: If most of probabilities are close to 1, that means model is very confident in its predictions. This could suggest that model is well-trained and that features are effective at predicting sentiment.
2. **Low confidence**: If most of probabilities are closer to 0.5 (in a binary classification problem) or spread out evenly (in a

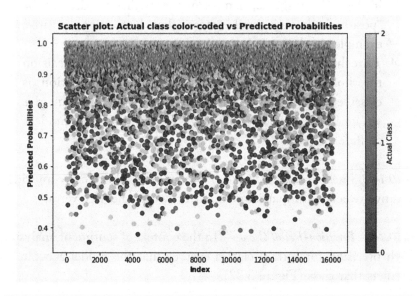

Figure 3.30   Scatter plot of actual vs predicted probabilities.

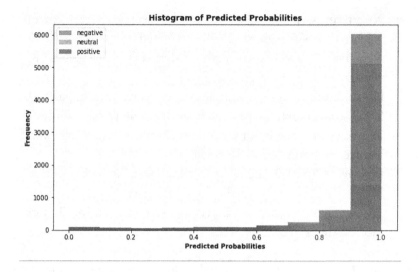

Figure 3.31 Histogram of predicted probabilities.

multiclass classification problem), this suggests that model is less certain of its predictions. This could mean model is underfitting or that features may not be effective at predicting sentiment.

3. **Class imbalance**: If the probabilities are heavily skewed towards a particular class, it might indicate a class imbalance in data. For instance, if most of the predictions are for the "positive" class, it could mean that have more positive examples in dataset.

4. **Threshold determination**: If to adjust the decision threshold (the probability above which a prediction is classified as a positive instance), the histogram can give an idea of a good value to pick based on the model's confidence levels.

3.10.14.3 Visualize the Model Performance For checking confusing positive tweets with neutral or negative tweets and vice versa.

3.10.14.4 Precision-Recall Curves In the context of sentiment analysis, high precision for certain classes ensures that the model accurately predicts that class (Figure 3.32).

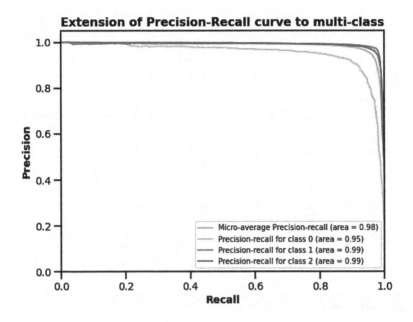

Figure 3.32 Extension of precision-recall curve to multi-class.

3.10.14.5 ROC Curves: To See the Changes in the Classification Threshold This is important in sentiment analysis if researchers want to minimize false positives or false negatives (Figure 3.33).

3.10.14.6 Sentiment Score The sentiment contains four scores: neg, neu, Pos, and compound (Table 3.8).

1. **Neg score:** The negative sentiment score ranges from 0 to 1, where 0 means no negative sentiment and 1 means extremely negative sentiment.
2. **Neu score:** The neutral sentiment score ranges from 0 to 1, where 0 means no neutral sentiment and 1 means extremely neutral sentiment.
3. **Pos score:** The positive sentiment score ranges from 0 to 1, where 0 means no positive sentiment and 1 means extremely positive sentiment.
4. **Compound score:** The overall sentiment score ranges from –1 to 1, where –1 means extremely negative sentiment, 0 means neutral sentiment, and 1 means extremely positive sentiment.

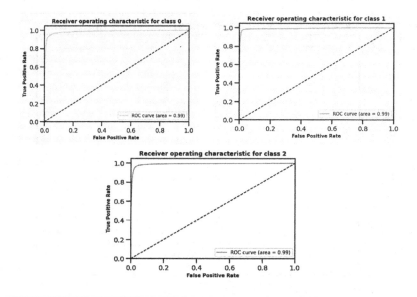

Figure 3.33   ROC curves: To see the changes in the classification threshold.

Table 3.8   Sentiment Score

```
# Trains a sentiment analysis model using the NLTK library:
import nltk
nltk. download('vader_lexicon')
from nltk. sentiment.vader import SentimentIntensityAnalyzer
# Load the sentiment analysis model
sentiment_model = SentimentIntensityAnalyzer()
# Example of using the model to get the sentiment score of a sentence
sentiment_scores = sentiment_model. polarity_scores ("This is a positive sentence.")
# Print the sentiment score
print(sentiment_scores)
{ 'neg': 0.0, # Negative sentiment score
'neu': 0.29, # Neutral sentiment score
'pos': 0.71, # Positive sentiment score
'compound': 0.5994} # Compound sentiment score
```

Anyone can use these scores to analyse the sentiment of the text. For instance, if the compound score is closer to 1, it means the text has a more positive sentiment, while if the compound score is closer to –1, it means the text has a more negative sentiment. If the compound score is close to 0, it means the text has a neutral sentiment.

Sentiment analysis to decentralized trading market prediction may not be straightforward, as they are two different domains with different types of data. But for this project as a book chapter, in terms of title, there is a connection; both have similarities, that is social media posts related to both the stock tweet and cryptocurrencies/blockchain.

To connect these results to the decentralized prediction markets by using AI, and ML, you can potentially use the insights gained from the analysis of the two datasets to inform the development of a decentralized prediction market platform.

The outcomes from the sentiment analysis for stock tweets and the combined Market Indicator Data ML model may be utilized to forecast the future performance of the trading market utilizing AI and ML approaches.

3.11 The Model (ML and AI) Integrated into Blockchain Technology

Blockchain technology is often used to build secure, decentralized systems for transactions and record-keeping, rather than for forecasting or modelling purposes. While it may be possible to build a decentralized trading market using blockchain technology, doing so would necessitate a deeper comprehension of the unique requirements and difficulties of such a system, as well as the proper design decisions for a blockchain-based solution. The smart contract can be created to take data from the trading market, preprocess it, and then use the ML model to forecast future events. The predictions can then be stored on the blockchain and made available to traders in a decentralized and secure manner. In addition, the decision can be made in another way: ML and AI models can be converted into smart contracts in the blockchain network, their recorded data will be saved in the network, and then traders or investors can predict the outcomes (Figure 3.34).

3.11.1 Process Flowchart of Blockchain Model Building

1. Transform the model into a smart contract: The model must be transformed into a smart contract to be incorporated into the blockchain network. A self-executing piece of code known as a smart contract can be installed on a blockchain

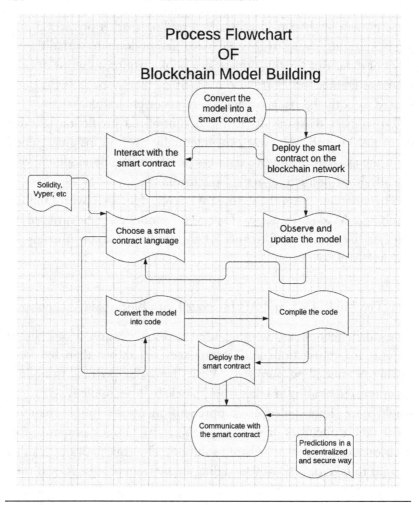

Figure 3.34 Process flowchart of Blockchain model building.

network. The predicted output should be returned by the smart contract, which should use the input data as its parameters.

2. Implementing the smart contract requires the smart contract code to be present on the blockchain network. You can use a blockchain platform like Ethereum, Hyperledger, etc., to implement the contract. Once you deploy the contract, every node in the network will have a copy of it, making it a permanent component of the blockchain network.

3. Communicate with the smart contract: To use the smart contract to predict the future, you must exchange

information with it. You can do this by sending a transaction containing the input data to the contract. The contract will then execute the code and return the desired outcome. A secure and decentralized prediction is made possible by the contract's setup on a decentralized network.

4. Observe the data and make model revisions: Performance can be assessed after the contract has gone into effect. Using information from the blockchain network, the model might be improved.

5. Choose a programming language that can create smart contracts from among Solidity, Vyper, and other options. Choose a language that satisfies your requirements and is compatible with your blockchain platform.

6. Model-to-code: Conversion is required so that the smart contract can make use of the ML model. Typically, this code will contain the input and output parameters for the model as well as any necessary functions for data processing.

7. Compile the code: Before the code can be executed on the blockchain, it needs to be changed into byte code. Depending on the smart contract language you select, the exact compilation procedure will change.

8. Deploy the smart contract: The smart contract can be posted to the blockchain network when the code has been compiled. To do this, a transaction that contains the byte code and any required metadata must typically be sent to the network.

9. Communicate with the smart contract: Following deployment, you can interface with the smart contract to make predictions in a decentralized and safe manner.

Solidity is a popular language for writing smart contracts on the Ethereum blockchain. To convert your ML model into a smart contract written in Solidity, you will need to do the following steps:

The most widely used programming language for creating smart contracts on the Ethereum blockchain is called Solidity.

ML model into a Solidity smart contract:

• Create the smart contract's Solidity code, which will contain your ML model. It will be necessary for this contract to have functions for accepting data and producing predictions using that data.

- Transform your ML model into code that the Ethereum Virtual Machine can execute (EVM). This may entail compiling your code into EVM byte code using a program like Web3.py or Brownie.
- Deploy the smart contract and any other dependencies, such as libraries or other contracts, to the Ethereum network.
- Send data to the smart contract and get predictions by interacting with it. This can be done using a tool such as Web3.py to interact with the contract's functions.

Figure 3.35 shows a flow chart for decentralized trading market prediction using ML, AI, and blockchain [57–59].

3.11.2 Basic Example of a Blockchain in Python

Each block in the chain has a complete picture in the output. The following characteristics for each block on the blockchain are printed (Table 3.9).

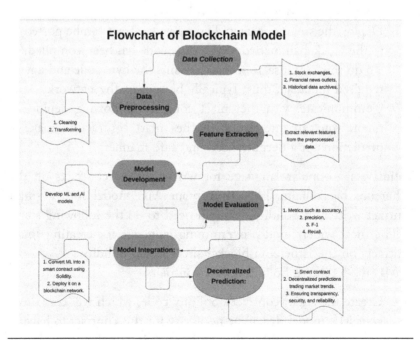

Figure 3.35 Flowchart of Blockchain model.

Table 3.9 Basic Example of Blockchain in Python

INDEX	TIMESTAMP	TRANSACTION	PREVIOUS HASH	NONCE	HASH
0	1685767837	Genesis Block	0	0	7e9e63be2d76758c7cf6fb02777ab7653-ca895d047f0b752b85bce4f161427db
1	1685767837	Transaction 1	7e9e63be2d76758c7cf6fb02777ab7653-ca895d047f0b752b85bce4f161427db	127	0050cf46107bb16a243bd546f1329cd772083d836d36531-da59fdcf1d5f463e9
2	1685767837	Transaction 2	0050cf46107bb16a243bd546f1329c-d772083d836d36531da59fdcf1d5f463e9	10	00cd657c4b4be69d8882a1ba821e65e23a5b2-def49c040d0f731cab94f8810f
3	1685767837	Transaction 3	00cd657c4b4be69d8882a1-ba821e65e23a5b2def49c040d0f731-cab94f8810f	36	005e0977d6e13401dc205a3c7fe77597ae496f719-f2380e57606d8eea12c31

1. **Index**: This is a unique identifier for each block in the chain. It typically starts at 0 for the genesis block and increments by 1 for each new block.
2. **Timestamp**: This is the time at which the block was created. It's typically represented as the number of seconds since the Unix Epoch (00:00:00 UTC on 1 January 1970).
3. **Transaction**: It is a string representing the transactions that are included in the block. In a real blockchain, this would usually be a list of actual transactions.
4. **Previous Hash**: It is the hash of the previous block in the chain. It's one of the key features of a blockchain that provides security and immutability. Any change in a block will change its hash, and since the hash is included in the next block, a change to any block would also change all subsequent blocks in the chain.
5. **Nonce**: This a number that has been found to make the hash of the block satisfy the difficulty level set for the blockchain. This is part of the proof-of-work algorithm used in many blockchains.
6. **Hash**: This is the hash of the block, calculated based on its contents (index, timestamp, transactions, previous hash, and nonce). A hash is a fixed-length string of characters that uniquely represents the data was hashed.

3.11.3 Blockchain Model in Python

Linear regression as an ML model has been made at this part for converting to Blockchain Model. And the observation was made during building the model in Python. The results of converting ML to Blockchain in Python have been mentioned below:

Linear regression algorithm model result:

RMSE: 0.17802817710736496

R^2 score: 0.5422962115201089

Previous ML part, there were 4 ML models being built and shown RMSE and R square scores. But in this part, a simple linear regression algorithm model has been made again for blockchain model because the blockchain model is very challenging for prediction. Just approaching manner implemented on linear regression ML to blockchain model for prediction. Here, has been used the same market indicator dataset for the ML and Blockchain model and shown the below result (Figures 3.36 and 3.37).

```
In [18]: # Train the model
         model = LinearRegression()
         model.fit(X_train, y_train)

         # Create a new blockchain instance
         my_blockchain = Blockchain()

         # Make a prediction and store it in the blockchain
         prediction = my_blockchain.make_prediction(user_input)
         print(f"Prediction: {prediction}")

         # Check the blockchain contents
         print("Blockchain contents:")
         for block in my_blockchain.chain:
             print(block.transactions)

Prediction: 2915.2156820294663
Blockchain contents:
Genesis Block
{'user_input': array([126.262543]), 'prediction': 2915.2156820294663}
```

Figure 3.36   Implementing a linear regression model and storing predictions in a Blockchain.

```
In [19]: import warnings
         from sklearn.metrics import mean_squared_error

         # Suppress warning messages
         warnings.filterwarnings('ignore', category=UserWarning)

         # Make predictions and store them in the blockchain
         my_blockchain = Blockchain()
         for index, row in df.iterrows():
             user_input = row['TRADING_VOLUME']
             prediction = my_blockchain.make_prediction(user_input)

         # Get the predicted values from the blockchain
         predicted_values = [block.transactions['prediction'] for block in my_blockchain.chain[1:]]

         # Get the actual values from the dataset
         actual_values = df['TRADING_VOLUME'].values

         # Calculate the RMSE
         rmse = mean_squared_error(actual_values, predicted_values, squared=False)
         print(f"RMSE: {rmse}")

         # Reset warning settings to default
         warnings.resetwarnings()

RMSE: 2816.8960812387363
```

Figure 3.37   Calculating the root mean square error (RMSE) of Blockchain-stored predictions against actual values.

The RMSE value of the blockchain model is a measure of how well the model can predict the TRADING_VOLUME values for the dataset. The lower the RMSE value, the better the performance of the model. Therefore, A high RMSE value indicates that the predictions made by the blockchain model are not accurate. In other words, the model is not performing well in predicting the decentralized trading market based on the input variables. A lower RMSE value would indicate better predictive performance. Good results depend on the dataset. Feature selection, cleaning pre-process technique, missing value, outlier removal have been applied, and so on. fill in the missing values with the mean or median of the column or use more advanced techniques like interpolation or imputation. The result has been shown from a selected secondary dataset is a small size and does not even have many attributes. A large volume and diverse source of datasets help to get a good result. Just a simple process of converting ML to the blockchain has been applied to come up with a conclusion and to see how it works. The process was correct, but no good results were shown because of the dataset size and fewer attributes. But good outcomes are possible with large quantities and many attributes.

In Python, a set of code makes predictions using the method of the Blockchain class, and then gets the predicted values from the blockchain and the actual values from the dataset. It then calculates the RMSE using the mean_squared_error function and prints the result. The prediction value is shown as 2915.2156820294663, which is the predicted value for the user input value of 126.262543. The Blockchain contents output shows that the blockchain had one block with the message "Genesis Block", indicating that this is the first block in the chain and a second block with the transaction details for the prediction.

In decentralized trading market prediction using ML and blockchain models, by using a blockchain, the predictions are stored in a decentralized and immutable manner, which can help ensure the integrity and transparency of the prediction system.

The prediction value 2915.2156820294663 is for the "TRADING_VOLUME" attribute only since the user_input value used to make the prediction was 126.262543, which corresponds to the normalized "TRADING_VOLUME" value. The amount and volume of the datasets are modest. The quality and quantity of the data, the ML methods and methodologies chosen, and the success of the blockchain implementation all play a role in whether a decent result can be obtained utilizing this dataset. It is crucial to test out various models and methods, as well as to continuously monitor and enhance the system's performance.

3.11.4 Challenges and Limitations of Using AI and Blockchain Technology in Trading Strategies

First, one of the challenges is related to the accuracy and reliability of AI algorithms. According to Han et al. [59], while AI can provide valuable insights and predictions, it is not infallible and may make mistakes or produce inaccurate results. Additionally, AI algorithms require significant amounts of historical data to be trained properly, which can be a challenge for newer markets or assets with limited data available [60].

Secondly, blockchain technology also presents its own set of challenges. One of the main limitations is the scalability of the blockchain network, as it can become congested and slow during periods of high transaction volume [61]. This can potentially impact the efficiency of trading strategies that rely on quick execution times.

Another challenge is the lack of regulation and oversight in the blockchain space, which can make it difficult to assess the credibility and legitimacy of certain blockchain-based trading platforms and assets [62]. Additionally, the decentralized nature of blockchain can also make it challenging to identify and address security vulnerabilities and risks.

Eventually, the challenges and limitations of using AI and blockchain technology in trading strategies include the accuracy and reliability of AI algorithms, the scalability and regulation of blockchain technology, and potential security vulnerabilities. It is important for traders to carefully evaluate these factors when incorporating these technologies into their trading strategies [63].

3.11.5 Future Research Directions and Opportunities

There are several potential future research directions and opportunities for using ML, AI, and blockchain technology to predict trading market strategy. One opportunity is to develop hybrid approaches that combine multiple technologies, such as using AI to analyse historical trading data and make predictions, while blockchain is used to securely execute trades and ensure transparency and trust in the process [64].

Another area of potential research is to explore the use of reinforcement learning algorithms, which can learn and adapt to changing market conditions in real-time, potentially improving the accuracy and effectiveness of trading strategies [65]. Additionally, with the rise of blockchain technology, there is an opportunity to create DAOs that use AI and ML algorithms to make decisions and execute trades on behalf of members, potentially leading to more democratic and decentralized financial systems [66].

However, as with any technology, there are also ethical considerations to consider when using ML, AI, and blockchain technology in trading strategies. Future research could explore the ethical implications of these technologies, such as the potential for bias and discrimination, and how to mitigate these risks [14].

3.12 Conclusion

ML models, including linear regression, decision trees, random forest, and XGBoost, have proven effective in predicting trading volume in decentralized markets. These models can be used to develop trading strategies and back-test them on historical data to evaluate performance. The integration of blockchain technology with AI and ML models supports secure, transparent, and efficient trading market predictions.

The XGBoost ML model outperformed others, with an R square value of .99 and RMSE value of .05. Logistic regression was used for sentiment analysis, comparing sentiment scores with a standard to aid in trading decisions. The research also outlines the integration of AI and ML models with blockchain technology, using smart contracts and deploying new blockchain models in Python. This study provides valuable insights into the development of trading strategies in decentralized markets using AI, ML, and blockchain technology, offering stakeholders data-driven decision-making tools and addressing potential biases and risks in the trading process. By harnessing these technologies, more secure, transparent, and efficient trading market predictions can be achieved in decentralized marketplaces.

3.13 List of Key Terms and Definitions

- **Artificial Intelligence (AI)**: The ability of machines to replicate human intelligence and cognitive processes through simulation.
- **Blockchain Technology**: A decentralized, distributed ledger that allows secure and transparent record-keeping of transactions.
- **Decentralized Prediction Markets**: Markets that use blockchain technology to allow anyone to participate in trading assets, based on their predictions of future events.
- **Machine Learning (ML)**: A subset of AI that involves using mathematical models and algorithms to enable machines to learn from data and improve their performance over time.
- **Prediction Models**: Models that use data to make predictions about future events or outcomes.
- **Smart Contracts**: Self-executing contracts that contain the terms and conditions of an agreement between parties and are stored on a blockchain.
- **Tokenization**: The process of converting assets into "tokens" that can be traded on a blockchain.
- **Probability Theory**: The branch of mathematics that deals with the quantification of uncertainty and the likelihood of events occurring.
- **Regression Analysis**: A statistical method used to examine the relationship between a dependent variable and one or more independent variables.
- **Decision Tree**: A tree-shaped diagram used to represent decisions and their possible consequences.
- **Supervised Learning**: A type of ML algorithm in which the machine is trained on labelled data to predict future outcomes.
- **Unsupervised Learning**: A type of ML algorithm in which the machine is trained on unlabelled data to identify patterns and structures.
- **Reinforcement Learning**: A type of ML algorithm in which the machine learns through trial-and-error to maximize a reward or minimize a penalty.

- **Cross-Validation**: A method used to evaluate the performance of an ML model by dividing the dataset into training and validation sets.
- **Feature Selection and Engineering**: The process of selecting the most relevant features from a dataset and engineering new features that improve the predictive power of the ML model.
- **Market Indicators**: Variables that are used to measure the performance of a market or asset.
- **Conceptual Framework**: A visual or written representation of the key concepts, variables, and relationships of a research project.
- **Root Mean Square Error (RMSE)**: A measure of the differences between predicted and observed values.
- **Performance Evaluation**: The process of measuring the effectiveness of an ML model.
- **Sentiment Analysis**: A type of ML algorithm that is used to analyse and evaluate the emotions expressed in a piece of text, such as social media like tweeter posts or customer reviews.

References

[1] Rajkomar, A., Dean, J., & Kohane, I. (2019). Machine learning in medicine. New England Journal of Medicine, 380(14), 1347–1358.

[2] Cao, J., & Zhang, S. (2020). Financial big data and artificial intelligence. Journal of Financial Data Science, 2(2), 1–14.

[3] Behnke, S., & Karakas, E. (2020). Self-driving cars: A review of recent developments and ethical considerations. Business Horizons, 63(6), 747–761.

[4] Baker, R. S. (2019). The impact of artificial intelligence on education: A review of the literature. International Journal of Artificial Intelligence in Education, 29(3), 614–650.

[5] Floridi, L. (2019). AI ethics: The birth of a research field. Philosophy & Technology, 32(4), 611–616.

[6] Jobin, A., Ienca, M., & Vayena, E. (2019). The global landscape of AI ethics guidelines. Nature Machine Intelligence, 1(9), 389–399.

[7] Zohar, A. (2015). Bitcoin: Under the hood. Communications of the ACM, 58(9), 104–113.

[8] Kshetri, N. (2018). Blockchain's roles in meeting key supply chain management objectives. International Journal of Information Management, 39, 80–89.

[9] Szabo, N. (1997). Formalizing and securing relationships on public networks. First Monday, 2(9). 10.5210/fm.v2i9.548

[10] Buterin, V. (2014). A next-generation smart contract and decentralized application platform. Ethereum.

[11] Swan, M. (2015). Blockchain: blueprint for a new economy. O'Reilly Media, Inc.

[12] Cocco, L., Pinna, A., & Marchesi, M. (2019). Blockchain interoperability: Challenges and opportunities. Future Generation Computer Systems, 97, 82–92.

[13] Crosby, M., Pattanayak, P., Verma, S., & Kalyanaraman, V. (2016). Blockchain technology: Beyond Bitcoin. Applied Innovation, 2(6-10), 71–81.

[14] Zheng, Z., Xie, S., Dai, H.N., Chen, X., & Wang, H. (2018). Blockchain challenges and opportunities: A survey. International Journal of Web and Grid Services, 14(4), 352–375.

[15] Li, X., Jiang, P., Chen, T., Luo, X. & Wen, Q. (2020). A survey on the security of blockchain systems. Future Generation Computer Systems 107, 841–853.

[16] Jiang, R., Song, X., Huang, D., Song, X., Xia, T., Cai, Z., Wang, Z., Kim, K.S., & Shibasaki, R. (2019, July). Deepurbanevent: A system for predicting citywide crowd dynamics at big events. In Proceedings of the 25th ACM SIGKDD international conference on knowledge discovery & data mining (pp. 2114–2122).

[17] Narayanan, A., Bonneau, J., Felten, E., Miller, A., & Goldfeder, S. (2016). Bitcoin and cryptocurrency technologies: A comprehensive introduction. Princeton: Princeton University Press.

[18] IBM. (2022). IBM Watson Health. Retrieved from https://www.ibm.com/watson-health

[19] Li, Z., Liang, X., & Li, J. (2019). Blockchain and AI: Complementary technologies for supply chain management. Supply Chain Management Review, 23(1), 8–13.

[20] Hanson, R. (2013). Combinatorial information markets for the prediction of scientific and technological advances. In The future of prediction (pp. 113–130). Springer, Berlin, Heidelberg.

[21] "Marketplaces Architecture in ocean protocol" (2020) Ocean protocol, 20 September. Available at: https://blog.oceanprotocol.com/ocean-market-an-open-source-community-marketplace-for-data-4b99bedacdc3 (Accessed: March 17, 2023).

[22] Gnosis. (2022). About Gnosis DAO [online]. Available at: https://www.gnosis.io/about(Accessed: January 16, 2024).

[23] *AI and Blockchain are decentralized in marketing strategy* (2017) YouTube. YouTube. Available at: https://www.youtube.com/@CodingTech (Accessed: March 18, 2023).

[24] Augur (2021). How Augur works. Retrieved from https://www.augur.net/how-it-works/

[25] Hastie, T., Tibshirani, R., & Friedman, J. (2009). The elements of statistical learning: Data mining, inference, and prediction. Springer Science & Business Media.

[26] Zhang, G., & Qi, M. (2005). Neural network forecasting for seasonal and trend time series. European Journal of Operational Research, 160(2), 501–514.

[27] Bernardo, J. M., & Smith, A. F. (2000). Bayesian theory (Vol. 405). John Wiley & Sons.

[28] Kahneman, D., & Tversky, A. (1979). Prospect theory: An analysis of decision under risk. Econometrica, 47(2), 263–291.

[29] Chatterjee, S., & Hadi, A. S. (2015). Regression analysis by example. John Wiley & Sons.

[30] James, G., Witten, D., Hastie, T., & Tibshirani, R. (2013). An introduction to statistical learning. Springer.

[31] Sharpe, W. F. (1966). The Sharpe ratio. The Journal of Portfolio Management, 42–46.

[32] Markowitz, H. (1952). Portfolio selection. The Journal of Finance, 7(1), 77–91.

[33] Bottou, L. (2010). Stochastic gradient descent. In Large scale machine learning (pp. 177–186). Cambridge University Press.

[34] Murphy, K. P. (2012). Machine learning: A probabilistic perspective. MIT Press.

[35] Géron, A. (2019). Hands-on machine learning with scikit-learn, keras, and tensorflow: Concepts, tools, and techniques to build intelligent systems. O'Reilly Media.

[36] Zhang, Y., & Kwok, J. T. (2017). Stock price prediction via discovering multi-frequency trading patterns. In Proceedings of the 23rd ACM SIGKDD international conference on knowledge discovery and data mining (pp. 1147–1156). ACM.

[37] Khalil, K. (2020). FCA maps out 'inherent risks' in the growth of algo and AI-based trading, The TRADE. Available at: https://www.thetradenews.com/fca-maps-inherent-risks-growth-algo-ai-based-trading/ (Accessed: March 11, 2023).

[38] Oodles-User (2023). Solving supply chain management challenges with blockchain, blockchain. Oodles. Available at: https://blockchain.oodles.io/blog/solving-supply-chain-management-challenges-blockchain/ (Accessed: March 11, 2023).

[39] Augur (2018). The Augur Whitepaper. Available at: https://www.augur.net/whitepaper.pdf (Accessed: June 06, 2023).

[40] Buterin, V. (2014). The power of prediction markets. Ethereum. Available at: https://ethereum.org/en/blog/the-power-of-prediction-markets/ [Accessed (insert date accessed)].

[41] Cavicchioli, M. (2021). Fetch.AI, the Artificial Intelligence project on blockchain, the cryptonomist. Available at: https://en.cryptonomist.ch/2021/11/07/fetch-ai-theartificial-intelligence-project-on-blockchain/ (Accessed: March 18, 2023).

[42] *Probability* (no date) *Math is Fun*. Available at: https://www.mathsisfun.com/data/probability.html (Accessed: March 20, 2023).

[43] Song, Y. Y., & Lu, Y. (2015). Decision tree methods: Applications for classification and prediction. Shanghai Archives of Psychiatry, 27(2), 130–135. 10.11919/j.issn.1002-0829.215044

[44] Hanson, R. (2007). The policy analysis market (A thwarted experiment in the use of prediction markets for public policy). *Innovations: Technology, Governance, Globalization*, 2, 73–88. 10.1162/itgg.2007.2.3.73.

[45] Assetonchain, n.d. Blockchain tokenization services. Assetonchain. Available at: https://www.assetonchain.com/services/blockchain-tokenization/ [Accessed 22 March 2023].

[46] Machine learning for materials developments in metals additive manufacturing – Scientific figure on researchgate. Available from: https://www.researchgate.net/figure/The-calculated-root-mean-squared-error-for-six-different-models-fit-to-the-dataset-shown_fig2_347215664 [accessed 22 Mar 2023].

[47] Cody, R. (2013). Applied statistics and the SAS programming language (5th ed.). Pearson Education.

[48] Samani, K. (2017). A primer on decentralized prediction markets. Multicoin capital. Available at: https://multicoin.capital/2017/12/07/primer-decentralized-prediction-markets/ [Accessed (insert date accessed)].

[49] Aalborg, H.A., Molnár, P., & de Vries, J.E. (2019). What can explain the price, volatility and trading volume of Bitcoin? Finance Research Letters, 29, 255–265.

[50] Liu, C., Liu, X., Wu, Y., Wang, L., & Zhao, J. (2020). Correlation-based feature selection for machine learning in big data: A review. Journal of Big Data, 7(1), 1–24.

[51] Scherer, D., Müller, A., & Behnke, S. (2010). Evaluation of pooling operations in convolutional architectures for object recognition. In Proceedings of the International Conference on Artificial Neural Networks (ICANN) (pp. 92–101). 10.1007/978-3-642-15825-4_10.

[52] Burnham, K. P., & Anderson, D. R. (2002). Model selection and multimodel inference: A practical information-theoretic approach. Springer Science & Business Media.

[53] Shao, J. (1997). An asymptotic theory for linear model selection. Statistica Sinica, 7(2), 221–264.

[54] Gelman, A., Carlin, J. B., Stern, H. S., & Rubin, D. B. (2013). Bayesian data analysis (Vol. 2). Chapman and Hall/CRC.

[55] Liu, Y., Wang, X., & Jiang, S. (Jan. 2021). Application of artificial intelligence in stock market prediction: A review. IEEE Access, 9, 5106–5120.

[56] Dinh, T. V., Liu, R., Zhang, M., Chen, G., Liu, J., & Ooi, B. C. (Jul. 2019). Towards scalable blockchain systems: A systematic analysis of blockchain scaling solutions. ACM Computing Surveys, 52(4), 1–39.

[57] Szabo, N. (Dec. 2003). Smart contracts: Building blocks for digital markets. Extropy Journal, 16, 1–14.

[58] Gomber, P., Koch, J., & Siering, M. (2018). Blockchain in financial services: What is it good for?. In proceedings of the 51st Hawaii international conference on system sciences.

[59] Han, H., Zhang, Y., & Sun, X. (2019). A review of the applications of artificial intelligence in the securities industry. Journal of Intelligent & Fuzzy Systems, 37(1), 1–15.

[60] Taiwo, J. (2021). Blockchain and its discontents. In Handbook of blockchain and cryptocurrencies (pp. 375–405). Springer.

[61] Alharthi, A., & Zawadzki, P. (2021). Machine learning in finance: An overview. In Machine learning and data science applications in industry (pp. 205–229). Springer.

[62] Giot, P., & Petropoulos, A. (2020). Reinforcement learning in finance: A review. Journal of Financial Econometrics, 18(3), 349–383.

[63] Nadig, T. (2020). The pros and cons of AI in investing. Retrieved from https://www.etf.com/sections/features-and-news/pros-and-cons-ai-investing

[64] Grossman, J. (2020). Decentralized autonomous organizations (DAOs). In Handbook of blockchain and cryptocurrencies (pp. 509–539). Springer.

[65] Mittal, R. (2019). Ethical considerations in AI-based trading. Communications of the ACM, 62(11), 38–40.

[66] Crosby, M., Pattanayak, P., Verma, S., & Kalyanaraman, V. (2016). Blockchain technology: Beyond Bitcoin. Applied Innovation, 2(6-10), 71.

4

AI AND BLOCKCHAIN AS SUSTAINABLE TEACHING AND LEARNING TOOLS TO COPE WITH THE 4IR

MD AMINUL ISLAM AND NAAHI MUMTAJ RIHAN

4.1 Introduction

The serial digital mutation of industries and society comprehended as the Fourth Industrial Revolution (4IR) is fueled by the fusion of cutting-edge technologies like artificial intelligence (AI), robots, the internet of things, blockchain, etc. (Turing, 2023). It shows a reinvigorated age of economic and social change and builds on earlier industrial processes that have happened since the late 18th century (Figure 4.1).

The exponential rise of data, the outcome of progressive analytics and unctuous intelligence, the composition of new business proto-types made imaginable by digital technology, and the expanding

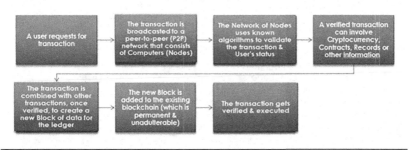

Figure 4.1 Blockchain process.

DOI: 10.1201/9781003162018-4

adoption of automation and robotics in distinct industries are some of the principal drivers of the 4IR (Figures 4.2 and 4.3).

AI can stimulate increased production and efficiency as well as the development of renewed, cutting-edge goods and services. The use of data and AI for virtuous purposes, the prospect of growing inequality and societal division, and the leverage of automation on occupation are all significant issues that are raised by this (Figure 4.4) (Turing, 2023).

The 4IR signifies a substantial change in the way we view technology and its effect on the future. To certify that technology aids everyone in society, it is vital that we are acquainted with its possibilities and challenges and band as a team. It is drastically altering not only the way we work and live but also how we educate our children. The way we teach and learn is evolving as digital technologies unfold more pervasive in our daily lives, and new skills are becoming more crucial for success in the job market of the future. The boosted focus on computational reflection and digital literacy in education is

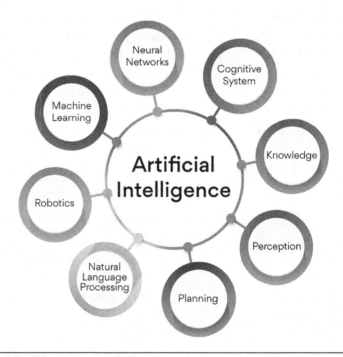

Figure 4.2   Different branches of AI.

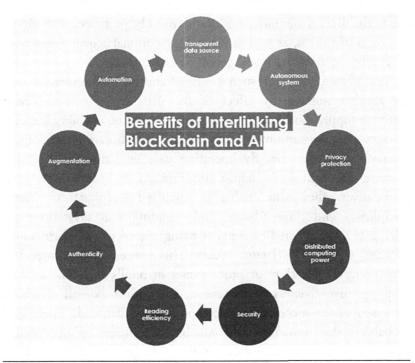

Figure 4.3   Interlink concept.

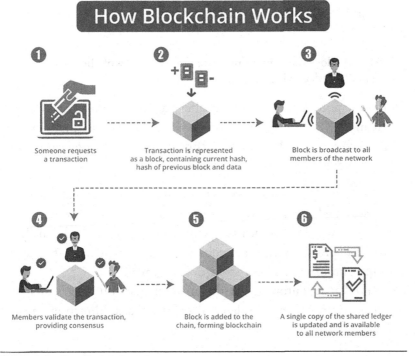

Figure 4.4   How Blockchain works.

one of the 4IR's major outcomes. Pupils must be proficient with digital tools and platforms, as well as have a foundational acquaintance with programming and data investigation (Takyar, 2023).

New learning formats, such as online and blended understanding are another noteworthy effect of the 4IR on education. These methods supply more adaptability and accessibility, enabling learners to access understanding materials and collaborate with others wherever they may be. By operating data analytics and adaptive learning algorithms to adjust information to user needs and prerogatives, they also enable personalized learning. An inter-disciplinary and project-based understanding that crystallizes on solving real-world predicaments is being placed on the curriculum because of the 4IR (Islam, 2022). This procedure promotes the blossoming of a variety of proficiencies in pupils including critical thinking, inventiveness, transmission, and teamwork—all of which are essential for success in the prospective job demand. There is a probability that numerous jobs will become ancient as automation and AI spread, and traditional educational designs may not be able to keep up with the rate of modification. This emphasizes the essence of lifelong wisdom and the elaboration of teaching to suit the directives of a world that is altering quickly.

4.2 AI and Blockchain in Education: An Overview of the Benefits and Challenges

AI has the potential to drastically alter education in many ways, such as streamlining administrative processes, reducing costs, and enhancing teaching and learning outcomes. The following are some potential advantages of AI in education:

- Learning Personalization: By customizing educational experiences and information to each student's specific needs and preferences, AI can facilitate personalized learning. Additionally, it can provide adaptive feedback and tests that help students recognize their areas of strength and growth.
- Intelligent Tutoring: AI-driven intelligent tutoring systems can offer students individualized feedback, direction, and

support as well as help them choose which subject, they need more help in. By automating a lot of administrative responsibilities including scheduling, grading, and data analysis, teachers can save time and money while also gaining more productivity. It can identify kids who may be in danger of falling behind or dropping out and offer them specialized treatments to help them succeed using predictive analytics.

- Digital Assistants: AI capabilities can provide teachers and students with on-demand support and guidance, resolving problems and disseminating information as needed. AI can help to improve language learning by listening to speech and providing real-time feedback on pronunciation, grammar, and vocabulary using natural language processing (NLP).
- Enhancing Accessibility: Various assistive technologies like text-to-speech and speech-to-text capability, AI can improve accessibility for students with disabilities (Ouyang and Jiao, 2021).
- Improving Research: Massive amounts of data may be quickly and successfully analyzed using AI to help academics uncover patterns and insights that may not be immediately apparent using more traditional methods (Figures 4.5 and 4.6) (Bhaskar, 2021).

The use of AI in education has a lot of potential advantages, but hazards and difficulties also need to be considered. Racism, privacy, and the impact of automation on the workforce are a few of the issues they raise. AI has the potential to change education and how we teach and learn through careful planning and implementation (Figure 4.7).

The following are some potential difficulties of AI in education:

- Bias: The impartiality of AI systems can only be ensured by the data used to train them. If the data used to train AI models is biased, the AI systems that are produced can likewise be prejudiced. As AI systems have the potential to reinforce preconceptions and sustain existing disparities, this raises issues in the realm of education. Privacy and Security: Such systems could gather and keep a lot of personal data, which raises questions about security and privacy. To protect sensitive data, it is crucial to make sure AI systems are created with the proper security features.

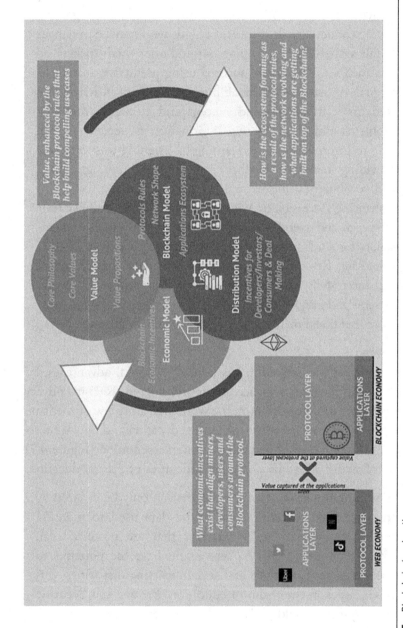

Figure 4.5 Blockchain in education.

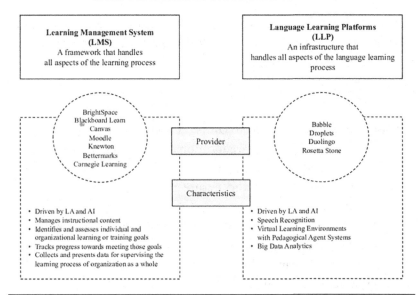

Figure 4.6   A new dynamic for EdTech in the age of pandemics.

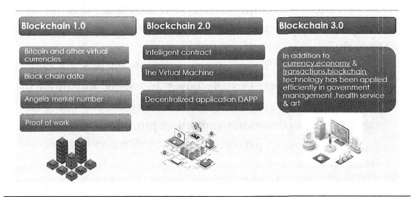

Figure 4.7   The development roadmap of Blockchain technology.

- Cost: Adopting AI systems may be difficult for educational institutions with limited financing since they can be expensive to design and install.
- Technical Complexity: Because AI systems can be sophisticated and challenging to comprehend, it may be challenging for administrators and teachers to use them effectively.
- Ethical Concerns: Ethical concerns are raised using AI in education, including concerns about accountability, transparency, and the impact of automation on the workforce.

Now, with a safe and open system for storing and exchanging educational data and certificates, blockchain technology (BCT) has the potential to revolutionize education. There are risks and issues with using blockchain in education.

Advantages

- Secure and Immutable Records: A secure and tamper-proof solution for keeping educational documents, such as diplomas and transcripts, is provided by BCT. This will help to prevent fraud and ensure that academic credentials are authentic.
- Increased Data Privacy: BCT enables people to manage their own data and share it with others only if they have a need to know, which can help preserve student data privacy and lower the risk of data breaches.
- Decentralized Learning Networks: Building decentralized learning networks with the aid of BCT will allow students and instructors to communicate with each other directly and free of intermediaries. Learning networks might consequently improve efficiency and cost.
- Smart Contracts: Smart contracts directly encode the terms of the contract between the buyer and seller into lines of code, and these contracts carry out their own execution. They can be used in education to automate procedures like student registration, tuition payments, and the validation of academic credentials.
- Micro-Credentials: Micro-credentials can be made using BCT to acknowledge abilities or accomplishments. By sharing their credentials with potential employers and preserving them on the blockchain, people are now able to describe their skills and expertise in more depth.
- Transparent and Trustworthy Learning Ecosystems: BCT can contribute to the development of trust and accountability among students, teachers, and institutions by establishing a transparent and reliable learning ecosystem and boosting overall educational standards (Figure 4.8) (Elayyan, 2021).

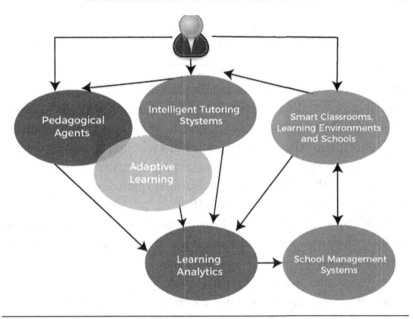

Figure 4.8   AI in the education system.

Issues with blockchain in education:

- Technical Complexity: It may be difficult for administrators and teachers to properly use BCT since it might be complex and difficult to understand.
- Interoperability: The variety of existing BC platforms can make it difficult for different systems to communicate with one another. When developing a uniform system for storing and exchanging educational data, it may provide difficulties.
- Scalability: Since BCT might be resource- and time-intensive, it might be challenging to scale it to serve extensive educational networks.
- Regulatory Concerns: Using BCT in education might provide regulatory and legal challenges when it comes to issues like data protection and intellectual property.
- Cost: The significant expenses connected with its development and deployment may make adopting BCT difficult for some educational institutions.

4.3 AI-Powered Personalized Learning: Customized Learning Experiences for Learners

The potential for individualized learning enabled by AI to totally revolutionize education is a rapidly expanding issue. Giving each student a personalized learning experience requires using AI algorithms to adapt instructional content to their unique requirements and preferences. AI-powered personalized learning can help students stay interested and motivated by presenting them with content that is pertinent to their interests and learning preferences, which can improve learning results (Figure 4.9).

With the use of AI-powered customized learning, the problem of student diversity may also be resolved. Teachers frequently find it difficult to satisfy the needs of students with various learning styles, experiences, and skills in typical classroom settings providing a personalized learning experience for each student that is suited to their individual needs and capabilities. AI-driven personalized learning has the potential to improve the learning process's efficacy and efficiency. AI-powered personalized learning can aid in streamlining and improving the learning process using computers to analyze student data and provide individualized suggestions (Mobius Consultants, 2023). A few examples of how AI may be used to create distinctive educational experiences are shown below:

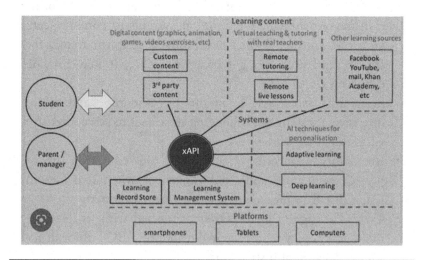

Figure 4.9 AI in education.

- Adaptive Learning: Adaptive learning systems driven by AI can examine student data to pinpoint areas in which they require additional assistance and suggest tailored learning resources. To ensure that each student is sufficiently challenged, these systems may also alter the content's level of difficulty in real time.
- Personalized Content: AI algorithms may help create individualized material for each student based on their learning preferences and interests. For instance, an AI-powered system may generate a student-specific reading list based on their preferred books and authors.
- Predictive Analytics: It can be used by AI to foresee a student's needs before they even materialize. AI algorithms can recognize which children are at risk of falling behind and provide specific actions to keep them on track by evaluating student data.
- Virtual Assistants: Virtual assistants powered by AI may reply to students' questions and provide them with personalized help, offering suggestions, and assisting them as they study. These assistants can modify their responses to fit the requirements of each learner.
- Gamification: Customized learning experiences that are enjoyable with gamification aided by AI. For instance, an AI-powered instructional game may modify the level of difficulty in accordance with the learner's advancement and provide specific criticism to assist them with improving.

One of the biggest challenges is the need for high-quality data. AI algorithms require access to accurate and trustworthy data to provide recommendations that are accurate. The possibility of bias, which can happen when algorithms are trained on biased data. AI-driven systems may dehumanize education by substituting automated systems for human contact (Davies et al., 2020).

4.4 Blockchain-Based Credentialing and Certification

A ground-breaking method for the examination and preservation of academic and professional credentials is BC-based credentialing. It entails leveraging BCT to produce impenetrable digital documents that are secure enough to share and verify (Islam et al., 2022).

The hassle of Verifying academic and professional qualifications has typically been a laborious and complicated process. Paper certificates and transcripts are issued by educational institutions and businesses, but these can be damaged, misplaced, or altered and need to be manually verified by employers, educational institutions, and other groups, which can be time-consuming and error-prone. By creating a decentralized, impervious-to-hacking method for validating and exchanging academic and professional credentials, blockchain-based credentialing seeks to streamline this procedure (Figure 4.10).

The implementation of blockchain-based credentialing is not without its challenges. One challenge is the need for uniformity. For digital credentials, there must be a format that is generally recognized. If blockchain-based credentialing is to be broadly used. The requirement for privacy presents another difficulty. Despite being more secure than traditional methods, Private data exchange is still required for blockchain-based credentialing which presents privacy issues.

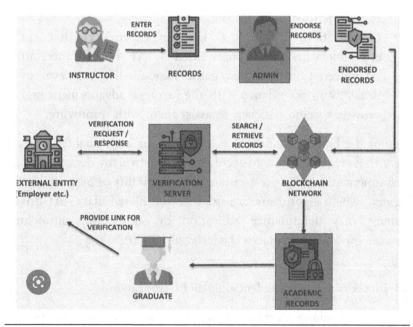

Figure 4.10 Blockchain-based academic records verification (Aamir et al., 2020).

Some of the main characteristics and advantages of blockchain-based certification are listed below:

- Decentralization: Blockchain-based certification is a decentralized system, which means that no single entity oversees the certification procedure. In its place, a network of nodes that verifies invalidated transactions maintains a distributed ledger.
- Immutability: Once a certificate is added to the blockchain, it cannot be modified or withdrawn. This increases the security of the certification process by removing the possibility of fraud.
- Transparency: Every transaction made on the blockchain is visible to everyone on the network and is transparent which is more transparent, which reduces the likelihood of errors and fraud.
- Efficiency: With no need for intermediaries like certification authorities, verification services, or other third-party institutions, blockchain-based certification is a more effective approach. which reduces the expenses involved with certification while also streamlining the process.
- Accessibility: The control over one's certification records is increased by utilizing blockchain-based certification. and can share them with future employers, institutions of higher learning, or other interested parties to improve their employability and mobility.

Creating secure, transparent records of diplomas and degrees using blockchains includes the following techniques:

- Creating digital records: BCT enables the production of digital records of educational credentials that are simple to access and distribute with the appropriate parties, like prospective employers or educational institutions. These documents provide information on the sort of degree or certificate received, the date of completion, and the name of the educational institution.
- Verifying records: A safe and impenetrable way for authenticating records of educational certificates is provided by BCT and eliminates the potential of fraudulent activities like phony certificates or diploma mills by keeping records on a decentralized ledger.

- Providing transparency: Because all transactions on the blockchain are visible, more transparency in the certification process is guaranteed, which everyone on the network may access. As everyone can view the same data, this lowers the chance of mistakes and fraud.
- Improving efficiency: Solutions for certification based on blockchain may be more efficient since they do away with the need for middlemen and expedite the entire certification procedure. This lowers the cost of the certification procedure and gives people the need to transfer their credentials between platforms or organizations more mobility.
- Enabling self-sovereign identity: A secure, decentralized system for managing digital identities can be made using BCT, giving users more control over their credentials and personal information. Such kind of self-sovereign identity can be used to confirm the identity of the certificate holder, enhancing the security and dependability of the certification process.

Using blockchain-based systems for educational qualifications is not without its difficulties, though. Systems may have trouble integrating and cooperating due to the inconsistent certification procedure, which is one of the key problems. The requirement for a trustworthy and secure digital identity system that can be used to confirm the identity of the certificate holder presents another difficulty. Some methods for using blockchain to build secure and open records of academic credentials are as below:

- Decentralization: BCT eliminates the need for a central authority to produce and manage certificates. Instead, certificates are kept on a distributed computer network, making it more challenging for fraudsters to alter the data or fabricate certificates. Decentralization enables people to manage their own certificates, providing them with greater freedom and flexibility.
- Security: A high level of security for educational credentials is provided by BCT where certificates are guaranteed to be secure and tamper-proof by using sophisticated cryptography and distributed consensus procedures. People can be assured that their certifications are safeguarded against unauthorized access or modification as a result.

- Interoperability: A universal standard for educational diplomas using BCT, many institutions and sectors will be able to distribute and recognize them easily by decreasing the complexity, this compatibility can help. and the expense of transferring certifications between various systems.

There are numerous instances of blockchain-based certification systems for education, such as the MIT Digital Diploma project, which issues and verifies digital degrees using BCT. Similarly, the Learning Machine startup has created a certification platform built on BCT that enables organizations to generate and distribute digital records of academic accomplishments.

4.5 AI-Powered Assessment and Evaluation

Intelligent assessment, commonly referred to as AI-powered assessment, is a developing field that employs AI to automate and enhance the assessment process in education. Massive amounts of data may be analyzed using AI to provide more accurate and useful analyses of student learning by utilizing algorithms of ML and NLP.

To achieve personalized learning for students, one should concentrate on educational adaptive learning technology, which uses intelligent methods to identify students' knowledge gaps and cognitive deficiencies, diagnose qualified steps for students, and thoroughly analyze data that can be designed using AI Bayesian formulas from (1) and (2).

$$P(A|B) = \frac{P(B|A) P(A)}{P(B)}, \tag{1}$$

$$OP(B) = P(B|A) P(A) + P(B|A) P(A), \tag{2}$$

The difference between the learner and the behavior of the curriculum education can be created and designed; for instance, students' physiological responses like heart rate, pulse, and skin temperature can detect students' learning behavior, or students' attention distribution can be detected through mouse and keyboard input, eye movements, etc., judging students' learning and interaction data based on eye tracking such as blinking and pupil dilation.

Also, the state of the learner can be diagnosed, and future development can be predicted, which can be imported through the following formulas:

$$L = \frac{1}{2}\sum_{i=1}^{mk}\left(Y^{(K)} - T_I\right)2, \tag{3}$$

$$\left(Y^{(K)} - TI\right)^2 = \frac{1}{2}\sum_{i=1}^{MK}(\partial i)^2, \tag{4}$$

where $\partial i = Y^{(K)} - T_I$ represents the difference between the i th element in any vector and the i th element

$$\frac{\partial L}{\partial W} = 0, \tag{5}$$

$$W - \partial \frac{\partial l}{\partial w} = 1. \tag{6}$$

∂ is the student learning rate and the step size of the weight, $(\partial l/\partial w)$ is the gradient;

The vector's ith element; the half value is added to make the subsequent derivation calculation easier; during the design phase, the student's responses can be incorporated into the regression equation to solve the various loss functions. Moreover, the network's weights W and W can be included in the design so that the reciprocal of L and W is 0, as indicated in equations (5) and (6).

It should be noted that it can be challenging to collect student physical characteristics accurately, making it necessary in many situations to estimate them. The estimate can be compared to the likelihood and incorporated into the regression equation, for example, to analyze letter pronunciation during reading and support personalized learning. Writing proficiency may also be assessed using a tablet for children who have trouble with it, allowing them to choose more suitable learning activities. AI may identify teaching techniques and processes that are more effective for students, categorize assignments based on feedback, pay more attention to the unique needs of each student, and examine learning trends to develop students' skills (Chen, 2022).

Following are some applications of AI-driven evaluation in the classroom:

- Automated grading: One of the most common uses is automated grading for AI-powered assessment. Essays, assignments, and tests can be analyzed by AI algorithms, which can then give students immediate feedback. This can help educators save time, lighten their load, and provide more instant feedback.

- Personalized learning: With the analysis of student data and the provision of individualized feedback and suggestions, AI-powered evaluation can also be utilized to personalize learning. These tools may identify pupils' areas of weakness and provide them with tailored feedback and tools to help them develop by analyzing student performance data (Ali and Abdel-Haq, 2021).

- Adaptive testing: The creation of adaptive tests, where the test's complexity adjusts based on the student's level of knowledge, is another application of AI-powered assessment which can offer a more accurate assessment of a student's abilities and can aid in lowering test anxiety.

- Learning analytics: Assessments enabled by AI can examine large volumes of student data to provide insights into student learning. Educators might find trends and patterns in student performance data that they can use to enhance teaching and learning.

- Curriculum design: AI can be used to analyze student data, create individualized learning plans, and adapt the curriculum to the needs of each individual student. Better learning results and greater student engagement may result from this.

However, integrating AI-powered assessment in education is not without its difficulties which include biases, availability of high-quality data to train the system, etc.

4.5.1 Plagiarism Detection

The detection of plagiarism in homework assignments can be aided by AI tools. Utilizing text-matching algorithms, which compare submitted work with a sizable database of existing texts, including submissions from other students, academic journals, books, and websites, is one popular strategy. These algorithms examine the degree of similarity between the submitted work and the references already in use, identifying any possible instances of plagiarism, i.e.,

Turnitin. Large Language Models (LLM), NLP, etc. methods can also be used to spot irregularities in writing style, suspicious patterns, or sudden changes in vocabulary. AI systems that have been trained on a large volume of text data can identify patterns that might be signs of plagiarism, like copying and pasting text from different sources without providing proper citations or paraphrasing. Machine learning algorithms are also used by some AI-powered plagiarism detection tools to continuously increase their accuracy. By studying past instances of plagiarism, they can spot new patterns or methods used by plagiarists, adjusting, and improving their detection abilities over time.

Some components of evaluation can be automated with the use of AI, improving the effectiveness, precision, and consistency of the procedure. Using AI to automate assessment can be done in the following ways:

- Grading: AI can be used to grade multiple-choice and short-answer questions automatically which also might expedite student feedback and free up instructors' time.
- Essay evaluation: It can be used to assess essays and other writing work. The writing's structure, content, and linguistic quality can all be examined to grade written work more quickly and consistently while leaving room for arbitrary criteria like creativity and originality.
- Feedback: Students can receive quick feedback from AI on their work, enabling them to identify their errors and potential improvement areas.

The possibility of bias in the algorithms employed for automated assessment is one of the key worries and algorithms may incorporate the programmers' prejudices or may have been trained on biased data which may lead to unjust judgments and impede their ability to learn.

Another challenge is the potential for cheating by submitting work that has been generated by AI or other tools. Students may try to take advantage of the algorithms to get a good score without really doing the work themselves (Ali and Abdel-Haq, 2021). But AI-based writing detection tools can resolve this issue.

Using AI to automate evaluation can be done in the following ways:

- Performance evaluation: AI can be used to assess a person's performance at work or in school. A detailed evaluation of a person's performance may be produced by AI algorithms by accessing data from a range of sources, such as productivity indicators, attendance, and work quality. Managers and teachers can save time by doing this, and assessments will be unbiased and data-driven solutions.
- Quality control: We may evaluate the quality of products and services that can be applied in manufacturing, for instance, to find flaws in goods that are being assembled. AI can be used in customer service to assess the effectiveness of interactions between clients and service providers, for example, by assessing the tone of emails or chat messages.
- Content evaluation: The worth and application of a piece of content may be determined using AI, including blog entries, videos, and social media updates. Content producers and marketers may find this helpful as they may utilize the comments to enhance their work and more successfully reach their target audience.
- Compliance evaluation: AI can help to assess adherence to laws and standards, for instance, in the healthcare industry to assess medical records and make sure they adhere to moral and ethical norms.

4.6 Blockchain-Based Decentralized Learning Networks

By building decentralized and secure platforms for students and instructors, BCT can completely transform the way learning networks work which makes it possible to build secure, open networks where users may communicate with one another directly and without middlemen.

- Peer-to-peer learning: Decentralized learning networks enable learners and educators to connect directly with each other, providing a peer-to-peer learning environment. So, regardless of their location or educational background, learners can gain from the knowledge of others in the network.

- Enhanced privacy: It offers improved privacy and security, safeguarding students' private data and making sure it is never shared without their permission.
- Open access: Open access to educational resources can be provided by decentralized learning networks and making it simpler for students to get the knowledge they need to succeed in their studies.
- Tamper-proof records: Using BCT, it is possible to produce tamper-proof records of academic success that anyone on the network may validate which offers a transparent and safe approach to track and validate academic credentials.

One of the biggest difficulties is the technology's intricacy, which can prevent some people from adopting it. The decentralized nature of the networks might make it difficult to maintain quality control and verify that all users are adhering to the same standards. Such technology has the potential to enable the establishment of decentralized learning networks, which could alter the way we learn and share knowledge (Alammary, 2019).

The following are some ways that BCT may make it possible to establish decentralized educational networks:

- Decentralized storage: With the aid of BCT, decentralized storage systems can be developed, allowing data to be kept on a network of computers rather than a single server which means that users can save and access learning materials from anywhere in the network, without relying on a centralized system.
- Immutable records: The ability to generate tamper-proof records of academic accomplishments that everyone in the network can verify is made possible by BCT and it ensures that the records are clear, correct, and cannot be changed or destroyed without authorization.
- Smart contracts: Smart contracts are carried out under the conditions of the parties' agreement and are represented by lines of code. BCT makes it possible to create smart contracts which can be used to automate some elements of education,

like tracking progress, confirming task completion, and rewarding students.

- Privacy and security: By BCT, consumers' personal information is protected and ensured to never be shared without their permission. Users may share their educational resources and successes on the network.

Decentralized learning networks could have the following advantages:

- Increased accessibility: For those who do not have access to conventional educational systems, decentralized learning networks can expand educational opportunities. It generally enables instructors to reach a larger audience as well as allows students to access educational resources from anywhere in the globe.
- Cost savings: By doing away with the need for costly infrastructure like actual classrooms, textbooks, and other resources, decentralized learning networks can lower the cost of education. This may increase the cost and accessibility of education for a wider group of students.
- Improved data security: Decentralized learning networks can make sure that sensitive data, such as students' personal information and academic records, are maintained on the blockchain in a safe and tamper-proof manner to preserve the privacy of students and avoid data breaches.

The following are some difficulties faced by decentralized learning networks:

- Technical complexity: The still developing and sophisticated BCT is the foundation for decentralized learning networks. Certain educational institutions and students may find it difficult to create and sustain a decentralized learning network.
- Adoption and scalability: Decentralized learning networks must be widely used, it can be difficult to persuade educational institutions, students, and other stakeholders to use this new technology. A network's ability to scale as its user base increases raises questions since it could have performance issues or congestion.
- Quality control: These might not have a central authority to control the caliber of exams and instructional materials. This

may result in problems like erroneous or biased content or low-quality assessments, which may ultimately damage the decentralized learning network's trustworthiness.

- Legal and regulatory challenges: These systems could run into legal and administrative problems, especially when it comes to data protection and intellectual property rights. BCT may potentially bring up legal and administrative issues pertaining to data ownership and control.
- Security risks: BCT is vulnerable to hacking and other security flaws. Both educational institutions and students must take precautions to make sure that their data is shielded from modification or illegal access.
- Connectivity and infrastructure: A stable and speedy Internet connection could be required for decentralized learning networks, which could not be offered everywhere. The absence of essential infrastructure, such as computers or electricity, may make it difficult for students to access decentralized learning networks.

To ensure that decentralized learning networks are available, safe, and efficient for all learners, it will be necessary for educational institutions, students, developers, and legislators to work together to overcome these obstacles (Alammary, 2019).

4.7 AI-Powered Content Creation and Curation

AI algorithms and techniques are used to create a variety of content, including text, photos, videos, and audio. In several areas, including education, this technology has the potential to transform content development. AI-powered content can quickly generate massive amounts of material which can be helpful in the educational setting, as teachers and instructional designers frequently need to produce a substantial amount of learning materials to satisfy the various demands of students (Figure 4.11).

The quality and accuracy of instructional content can both be increased with the use of AI-powered content development. AI can make sure that written text is clear and grammatically correct by employing NLP techniques. Data analysis and insight-providing

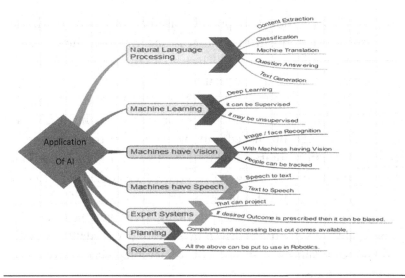

Figure 4.11 AI in education landscapes.

capabilities of AI-powered content production platforms can assist teachers and instructional designers in producing more effective teaching materials. Another benefit of AI-powered content creation is its ability to customize content to the needs and preferences of certain learners. AI may create specialized learning materials that are adapted to the individual needs of each student by analyzing data on learner behavior and performance.

AI-generated content can be unoriginal and uncreative. Although AI can produce content rapidly and effectively, it might not be able to do so in a way that is truly novel or ground-breaking. This might be a problem in the classroom, where students might gain from being exposed to fresh, original ideas. AI-powered curation uses algorithms and methods to automatically sort, arrange, and suggest content based on user interests and behavior. The way learning resources are chosen and provided might be completely changed by this technology.

Personalization of the learning experience through AI-powered curating is one of its primary advantages since AI algorithms can make content recommendations that are pertinent to a student's interests and learning objectives by examining data on learner behavior and preferences. As they are more likely to find the subject engaging and valuable, this can help students stay motivated and engaged.

AI-powered curation may provide insights into learner behavior and performance and can find patterns and trends in how learners interact with curated content, which they can then use to enhance the efficacy of the next curation initiatives.

Scalability is one of the key advantages of content creation powered by AI. In poor nations or for students who might not have access to high-quality educational resources, this can increase access to and affordability of education. The caliber of educational resources can be raised using AI in content creation. For instance, chatbots powered by AI can simulate discussions with learners, assisting them in honing and practicing their language abilities. Virtual reality (VR) settings driven by AI can give students dynamic, hands-on experiences that are not achievable with conventional instructional tools.

The requirement for human oversight and involvement presents another difficulty. The text produced by the algorithm must be reviewed and edited by human professionals. Critics are concerned that the use of AI tools in education could diminish students' capacity for critical and analytical thought throughout the learning process, these tools may unintentionally discourage students from utilizing critical thinking and problem-solving skills. The need for students to conduct research and exercise critical thinking is frequently diminished by AI tools that provide ready-made answers or solutions. This reliance on AI-generated responses may inhibit their ability to analyze complex problems, independently evaluate information, and generate original solutions. To provide students with a well-rounded education that will prepare them for the challenges of the future, it is necessary to establish a balance between the development of critical thinking skills and the use of AI tools. The ability of AI-powered material curation to save time for teachers and students is another prime advantage. Learners can rely on AI algorithms to suggest top-notch materials that are pertinent to their needs rather than spending hours searching for educational resources. Teachers can utilize AI-powered curation to find resources that are compatible with their learning goals, saving time and effort. The recommendations made by the algorithm may be biased if the data used to train the algorithm was skewed in any way which may expose students to information that is erroneous or biased. AI algorithms might offer helpful recommendations, but it is crucial to make sure that the suggested resources are precise, pertinent, and in line with learning goals (Little One The Jaipuria Preschool, 2023).

Changes in instructional design, teaching strategies, and teacher preparation are just a few of the ways that the use of AI in education will have a big impact on teaching. Here are a few examples of how AI might affect education:

- Personalized learning
- Adaptive learning
- Data-driven instruction
- Automated grading
- Intelligent tutoring systems

Now the application of AI in education does, however, present certain potential difficulties. For instance:

- Concerns about bias: AI algorithms' quality is based on the data that was used to train them. The recommendations or judgments that the algorithm makes may be prejudiced if the data utilized to train the algorithm is biased in some way. It's crucial to make sure AI is applied fairly, transparently, and with as little chance of prejudice as possible.
- The need for teacher training: Teachers must possess a specific level of technological knowledge and skills to deploy AI in the classroom. To successfully integrate AI into their teaching practices, teachers may require help and training.
- The risk of dehumanization: Aspects of education that can be automated by AI include grading and assessment. And while doing so can save time and effort, it is essential to use AI in a way that complements rather than displaces human participation and teaching.

Here are some ways that learning may be affected by AI:

- Personalized Learning
- Adaptive Learning
- Improved Student Engagement
- Accessibility
- Real-time Feedback

- Enhanced Collaboration
- Time Management

Yet, after all these discussed above there are some issues and difficulties with using AI in teaching also.

- Bias: Because AI algorithms have the potential to be prejudiced, different student groups may experience unequal learning outcomes.
- Dependence on Technology: Overreliance on technology among students may result in a dearth of creativity and critical thinking abilities.
- Privacy and Security: Since student data may be subject to hacking and other online dangers, the use of AI in education poses privacy and security concerns.
- Implementation: Infrastructure, support, and training must be heavily invested in for AI in education to be successful.
- Job Displacement: By using AI to automate some duties, the educational system may become unbalanced, and teachers' jobs may be lost (Ahmed and Ganapathy, 2021).

4.8 Case Studies: Examples of AI and Blockchain in Education, and Their Impact on Teaching and Learning

Below are some studies and illustrations of the use of blockchain and AI in education:

Smart Degrees in Malta: To introduce "Smart Degrees," the Malta government worked with Learning Machine, a BCT business. This approach safeguards academic credentials by storing them on a blockchain so they can't be changed or tampered with. In various universities in Malta, the system has been put in place, and students can access their credentials using a safe digital wallet.

Carnegie Mellon University's AI Tutor: An AI-powered teaching system that offers pupils individualized feedback and direction was developed by Carnegie Mellon University. The "ALEKS" system adapts to each student's learning style and speed, giving them content that is specifically designed to help them increase their knowledge and comprehension (Koedinger et al., 1997).

The University of Bahrain's Blockchain-based system: A blockchain-based system was put in place by the University of Bahrain to store and distribute student records and diplomas. Employers and other educational institutions may verify the data thanks to this method, which makes sure it can't be tampered with (JIbrel, 2019).

Also, the impact of AI on teaching and learning is seen in the following case studies:

Carnegie Learning: AI is used by Carnegie Learning to give pupils individualized learning opportunities. The technology adapts to the learning style and pace of each learner and uses machine-learning algorithms to spot places where they need more help. As a result, student outcomes have significantly improved; according to one study, students who use Carnegie Learning performed 23% better on examinations than those who used conventional techniques.

Coursera: An online learning platform called Coursera employs AI to customize each student's educational experience. Using machine learning techniques, the system evaluates each student's academic performance, then offers tailored suggestions for programs and resources. Higher engagement and completion rates have resulted from this; according to one study, students utilizing Coursera were 21% more likely to finish a course than those using conventional means.

Here are a few case studies that demonstrate how BCT affects teaching and learning:

ML: Blockchain-based digital credentials can be created and issued using software from Learning Machine. The Massachusetts Institute of Technology (MIT) is one of its customers, and it has been using ML technology to provide graduates with digital certificates since 2018. The verification process is accelerated, made safer, and made more visible using blockchain-based credentials.

Open University: Since 2015, The Open University in the United Kingdom has used blockchain to provide graduates with digital badges. The badges, which students can publish on social media and professional networking websites, signify the abilities and knowledge that they have acquired via their courses. Using blockchain-based badges has allowed the Open University to improve its standing and attract more students.

4.9 Challenges of AI and Blockchain: The Demerits of These in Teaching and Learning

The use of AI and blockchain in education could have certain negative effects, even though they have the potential to change teaching and learning. The following are some drawbacks of blockchain and AI in education:

- Lack of Human Interaction
- Bias and Inequality
- Cost
- Data Privacy and Security
- Complexity

It has become a common concern for education providers and teachers regarding LLM-based AI tools which can be used for cheating in assessments, homework, assignments, etc. Even some creative works like writing songs, poems, articles, paintings, and drawings. It can be used for primary-level idea generation but requires an adequate level of research, fine-tuning, and in-depth understanding to avoid paraphrasing and AI detection errors (IOE London Blog, 2023). ChatGPT can generate too generic answers to different types of questions but those are neither always correct nor high standards (Figure 4.12) (Code Today, 2022).

GPT and LLM models can scale up the problem-solving strategies faster and engage as a form of chat to formulate the ideas to narrow down (Tsai et al., 2023).

4.10 Conclusion

Personalized, decentralized, and secure learning networks can be created by integrating blockchain and AI into teaching and learning, which has the potential to completely change the way that education is provided. While blockchain-based credentialing and certification can offer a transparent and unchangeable record of learners' accomplishments, AI-powered personalized learning can adapt educational experiences to individual needs. Whereas blockchain can make it easier to create decentralized learning networks, certain components of evaluation and content creation can be automated

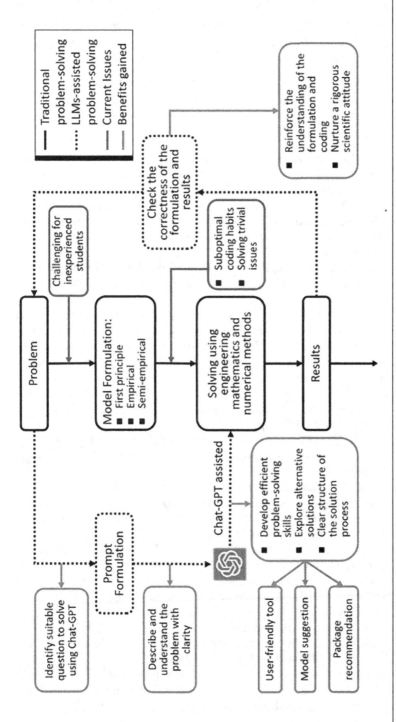

Figure 4.12 LLM-based approach for problem-solving.

with AI. Some of the challenges that must be solved include the need for appropriate standards and regulations as well as moral concerns with data privacy. Future studies should concentrate on examining the possible advantages and disadvantages of these technologies and creating best practices for their adoption in the classroom.

References

Aamir, M., Qureshi, R., Khan, F.A. and Huzaifa, M., 2020. Blockchain based academic records verification in smart cities. *Wireless Personal Communications*, 113, pp. 1397–1406.

Ahmed, A.A.A. and Ganapathy, A., 2021. Creation of automated content with embedded artificial intelligence: A study on learning management system for educational entrepreneurship. *Academy of Entrepreneurship Journal*, 27(3), pp. 1–10.

Alammary, A., Alhazmi, S., Almasri, M. and Gillani, S., 2019. Blockchain-based applications in education: A systematic review. *Applied Sciences*, 9(12), p. 2400.

Ali, M. and Abdel-Haq, M.K., 2021. Bibliographical analysis of artificial intelligence learning in Higher Education: Is the role of the human educator and educated a thing of the past?. In *Fostering Communication and Learning With Underutilized Technologies in Higher Education* (pp. 36–52). IGI Global.

Bhaskar, P., Tiwari, C.K. and Joshi, A., 2021. Blockchain in education management: Present and future applications. *Interactive Technology and Smart Education*, 18(1), pp. 1–17.

Chen, Y., 2022. The impact of artificial intelligence and block chain technology on the development of modern educational technology. *Mobile Information Systems*.

Davies, J.N., Verovko, M., Verovko, O. and Solomakha, I., 2020, June. Personalization of e-learning process using AI-powered chatbot integration. In *International Scientific-practical Conference* (pp. 209–216). Cham: Springer International Publishing.

IOE London Blog. 2023. What we should really be asking about ChatGPT et al. when it comes to educational assessment [Blog post]. UCL Institute of Education. Available at: https://blogs.ucl.ac.uk/ioe/2023/04/27/what-we-should-really-be-asking-about-chatgpt-et-al-when-it-comes-to-educational-assessment/. (Accessed 10 July 2023)

Islam, M.A., Sufian, M.A. and Sifat, M.H. 2022. AI, blockchain and self-sovereign identity in higher education. MRN Annual Conference 2022, UCL. doi: 10.13140/RG.2.2.29117.44008

Jibrel, 2019. Bahrain emerges as a block chain leader in the Middle East. Available at: https://cointelegraph.com/news/university-of-bahrain-issues-blockchain-based-diplomas-to-graduating-students (Accessed 1 March 2023)

Koedinger, K.R., Anderson, J.R., Hadley, W.H. and Mark, M.A., 1997. Intelligent tutoring goes to school in the big city. *International Journal of Artificial Intelligence in Education*, 8(1), pp. 30–43.

Little One The Jaipuria Preschool, 2023. *Artificial Intelligence in Education landscape.* Available at: http://www.jaipurialittleonejajmau.com/blogspot/ artificial-intelligence-in-education-landscape/ (Accessed 1 April 2023)

Ouyang, F. and Jiao, P., 2021. Artificial intelligence in education: The three paradigms. *Computers and Education: Artificial Intelligence*, 2, p. 100020.

Takyar, A., 2023. *Blockchain technology explained.* Available at: https://www. leewayhertz.com/blockchain-technology-explained/ (Accessed 1 April 2023)

Tsai, M.L., Ong, C.W. and Chen, C.L., 2023. Exploring the use of large language models (LLMs) in chemical engineering education: Building core course problem models with Chat-GPT. *Education for Chemical Engineers*, 44, pp. 71–95.

Turing, 2023. How do AI and blockchain technology complement each other? Available at: https://www.turing.com/kb/how-blockchain-and-ai-complement-each-other (Accessed 1 April 2023)

5

BLOCKCHAIN AND AI FOR HEALTHCARE

A Review of Opportunities and Challenges

NAZMUL HOQUE

5.1 Introduction

The healthcare industry has paid a lot of attention lately to blockchain (BC) and artificial intelligence (AI) based technology. In several areas, like speech recognition, facial recognition, and visual search, AI has already had considerable success. Histology tissue segmentation, brain magnetic resonance imaging (MRI) analysis, blood cell analysis, and many more deep learning-based medical imaging applications are just a few examples. The integration of AI, blockchain, and wearable technology can improve chronic disease management, moving from a hospital-centered model to a patient-centered one. However, organizing and analyzing data is a pressing issue, and blockchain can improve healthcare services by authorizing decentralized data sharing, protecting privacy, and ensuring reliable data management.

In contrast, the modern medical infrastructure faces serious security challenges when adopting smart healthcare systems. These systems offer several benefits but are vulnerable to cyber-attacks. Integrating blockchain technology with smart healthcare systems [1] can offer decentralization, immutability, transparency, and security. This chapter provides a comprehensive analysis of the emerging technologies, security attacks, and taxonomy of smart healthcare security solutions. This chapter also presents a case study for mitigating security attacks in smart healthcare systems and discusses the open issues and research challenges hindering the performance of the smart healthcare system.

DOI: 10.1201/9781003162018-5

By addressing these issues, This chapter seeks to help readers gain a better understanding of the security issues with the smart healthcare system and how blockchain and AI technologies may be used to reduce such risks. The integration of AI and blockchain-based technologies in healthcare offers significant opportunities to improve patient care, and this chapter discusses the various opportunities and challenges associated with this integration [2].

5.2 Integration of Blockchain and AI in Healthcare Supply Chains

The utilization of blockchain and AI technology to track the supply chain of pharmaceutical supplies and prevent counterfeits has gained significant traction in recent years.

5.2.1 Blockchain Technology in Healthcare Supply Chains

Blockchain's immutability ensures that once information about a pharmaceutical product is recorded on the blockchain, it cannot be altered without consensus from the network participants. This feature provides a tamper-proof system, that prevents counterfeiters from manipulating the supply chain data. By maintaining an unchangeable record of each transaction or transfer of a pharmaceutical product, blockchain creates a comprehensive and transparent view of its journey from the manufacturer to the end consumer. This transparency, as emphasized in my thoughts, enables stakeholders to track the provenance and movement of pharmaceutical supplies [3] at any given point in time.

Additionally, the integration of unique identifiers and cryptographic features with blockchain helps in verifying the authenticity of pharmaceutical supplies. Digital fingerprints, created through the combination of blockchain and unique identifiers like serial numbers, QR codes, or RFID tags, establish a secure and transparent way to track the movement of products throughout the supply chain. This verification process aids in preventing counterfeits and ensures that the medications are authentic and not compromised.

5.2.2 Synergy Between Blockchain and AI

The potential synergy between blockchain and AI technologies is enhancing supply chain management. AI can leverage the data stored on the blockchain to predict demand, prevent shortages, and identify instances of fraud or counterfeit products. By combining the strengths of blockchain's transparency and AI's predictive capabilities, the efficiency, quality, and safety within the pharmaceutical supply chain ecosystem can be significantly improved.

While these advancements are promising, acknowledge the challenges associated with implementing blockchain technology, such as interoperability and standardization issues. Collaboration and further research are crucial to overcome these hurdles and fully realize the potential of blockchain and AI in pharmaceutical supply chain management.

The utilization of blockchain and AI to track the supply chain of pharmaceutical supplies and prevent counterfeits brings forth immense benefits, including transparency, traceability, and enhanced security [4]. By recording and verifying the history of pharmaceutical goods on a tamper-proof blockchain, stakeholders can ensure data integrity, authenticate medications, and combat the growing issue of counterfeit products. However, addressing challenges and fostering collaboration remains key to harnessing the full potential of blockchain and AI technology in pharmaceutical supply chain management.

5.3 Blockchain and AI in Accessing and Sharing Patient Data

The integration of blockchain and AI in accessing and sharing patient data holds immense potential to revolutionize healthcare. This analysis delves into the vital aspects of security and privacy in patient data, the challenges in implementing blockchain and AI [5], and the future prospects of these technologies in the healthcare landscape.

5.3.1 Security and Privacy of Patient Data

Both security and privacy are critical considerations when accessing and sharing patient data. Blockchain technology provides a decentralized and tamper-proof ledger that enhances security [6]. Its immutability and distributed nature ensure data integrity, making it resistant to unauthorized modifications and breaches. Patient data stored on a blockchain can be securely accessed by authorized parties, promoting transparency and accountability.

Furthermore, AI algorithms play a crucial role in analyzing patient data, enabling the identification of patterns and trends that can improve patient care. However, the combination of AI and blockchain raises concerns regarding data privacy. To address this, blockchain can pseudonymize patient data, encrypting it while storing only unique identifiers on the ledger. This approach enhances patient privacy and minimizes the risk of unauthorized identification and tracking.

5.3.2 Challenges of Implementing Blockchain and AI in Healthcare

Despite the vast potential of blockchain and AI in healthcare, there are significant challenges to address. One primary hurdle is the lack of industry-wide standards for blockchain implementation, making interoperability between different organizations and systems complex. Additionally, the evolving regulatory landscape poses uncertainties that organizations must navigate when adopting these technologies.

Interoperability is further compounded by the diverse range of blockchain platforms available, necessitating careful consideration when selecting the most suitable platform for specific healthcare requirements. Moreover, ensuring patient education about the benefits, risks, and proper usage of blockchain technology is vital for fostering trust and widespread adoption.

Security risks associated with blockchain, such as potential vulnerabilities in smart contracts or distributed denial-of-service attacks, must be thoroughly evaluated and mitigated to ensure the integrity of patient data.

5.3.3 Future Prospects

Looking ahead, the future of blockchain and AI in healthcare is promising. These technologies can significantly improve the quality of care and make healthcare more affordable and patient-centric. Blockchain's ability to facilitate patient-controlled data ownership and AI's advanced diagnostic and treatment capabilities offer exciting possibilities.

Decentralized healthcare applications built on blockchain can empower patients to manage and control their health data securely [7]. This shift toward patient-centricity promotes data transparency and empowers individuals to make informed decisions about their healthcare.

Moreover, the integration of AI and blockchain can lead to the development of innovative diagnostic tools, treatment plans, and precision medicine approaches. By analyzing vast amounts of patient data, AI algorithms can enhance clinical decision-making, improve disease prediction, and drive personalized healthcare delivery.

To fully realize the potential of blockchain and AI in healthcare, collaboration among stakeholders, including healthcare providers, technology experts, regulators, and policymakers, is essential. Establishing robust standards, addressing regulatory challenges, and investing in education and research are critical steps toward widespread adoption and the seamless integration of these transformative technologies.

The integration of blockchain and AI in accessing and sharing patient data offers significant benefits for the healthcare industry. By addressing security and privacy concerns, overcoming implementation challenges, and leveraging the potential of these technologies, healthcare can be revolutionized, fostering improved patient outcomes and a more patient-centric approach to care.

5.4 Integration of Blockchain and AI in Healthcare Remote-Patient Monitoring

Blockchain and AI present unprecedented opportunities to transform healthcare. The integration of these technologies can revolutionize RPM systems, allowing clinicians to remotely monitor patients and deliver timely interventions [8]. Blockchain provides secure and tamper-proof storage of patient data, while AI analyzes this data to identify health patterns and enable appropriate actions.

5.4.1 Benefits of Integrating Blockchain and AI in RPM Systems

5.4.1.1 Improved Privacy and Security Blockchain ensures patient privacy by securely storing and sharing sensitive data. Its tamper-proof nature safeguards patient information, reducing the risk of unauthorized access or data loss.

5.4.1.2 Enhanced Efficiency and Accuracy of Data Analysis   AI algorithms analyze patient data collected from various sources, enabling efficient and accurate data analysis. This facilitates early identification of health problems, enabling timely interventions and better patient outcomes.

5.4.1.3 Cost Reduction in Healthcare   The integration of blockchain and AI in RPM systems has the potential to reduce healthcare costs. By identifying health issues early on, interventions can be implemented promptly, potentially preventing the progression of diseases and reducing the need for expensive treatments [9].

5.4.2 Challenges in Integrating Blockchain and AI in RPM Systems

5.4.2.1 Need for Sufficient Data   To fully leverage the potential of AI in RPM, a sufficient volume of high-quality patient data is essential. Access to diverse and comprehensive datasets is crucial for training accurate AI models.

5.4.2.2 Development of Advanced Algorithms   Continued research and development of AI algorithms are necessary to improve the accuracy and efficiency of data analysis in RPM systems. Ongoing advancements in machine learning and deep learning techniques are vital to extracting meaningful insights from patient data.

5.4.2.3 Regulatory Considerations   The integration of blockchain and AI in RPM systems necessitates a robust regulatory framework. Regulations must address data privacy, security, and ethical concerns to ensure the responsible and ethical use of patient data.

5.4.3 Future Prospects

The integration of blockchain and AI in RPM systems holds immense potential for revolutionizing healthcare. As the technologies continue to evolve, we can expect even more innovative applications [10]. To realize this potential, collaborative efforts among researchers, healthcare providers, regulators, and technology developers are crucial. By addressing the challenges and leveraging the benefits, we can embrace a future of patient-centric, cost-effective, and secure remote healthcare.

5.5 Blockchain and AI in Healthcare Insurance

The insurance industry is ripe for disruption. The traditional way of doing things is no longer sustainable, and insurers need to embrace new technologies in order to stay competitive. Blockchain and AI [11] are two of the most important technologies that will shape the future of insurance.

5.5.1 Problem Statement

The significant issue of healthcare insurance fraud, which results in substantial financial losses for the industry is enourmous. Traditional fraud detection methods are often inefficient and time-consuming, necessitating the exploration of new technologies such as blockchain and AI to address this challenge effectively.

5.5.2 Role of Blockchain and AI

The potential of blockchain as a secure and transparent data storage technology. By utilizing blockchain to record healthcare transactions, patterns of fraud can be identified and analyzed. AI, in turn, can leverage the data stored on the blockchain to perform sophisticated analysis and identify potential fraud cases.

5.5.3 Automation and Efficiency

The benefits of automation in fraud detection are outlined, allowing human resources to focus on other critical tasks, such as investigating suspected cases of fraud [12]. By employing blockchain and AI, the process of claim adjudication, risk assessment, and investigations can be automated, resulting in increased efficiency and accuracy.

5.5.4 Architecture of Fraud Detection System

The integration of a blockchain network, data repository, and AI-powered fraud detection engine forms the foundation of this system. These components work together to store, track, and analyze healthcare data for identifying potential fraud cases.

5.5.5 Future Prospects

Blockchain and AI hold significant potential for revolutionizing healthcare insurance fraud detection [13]. They foresee a more efficient, effective, and accurate fraud detection system by leveraging these technologies.

The benefits of using blockchain and AI in healthcare insurance fraud detection, such as automated claim adjudication, risk assessment, and investigations, further reinforce the potential of these technologies in combating fraud.

5.6 Future of Blockchain and AI in Healthcare

The convergence of blockchain and AI capabilities in the healthcare domain may bring a sea change in its landscape; unlocking fresh pathways for advancements and efficiency. This composition delves into the collaborative potential of these two cutting-edge technologies in navigating vital health-related challenges, including data security assurance, streamlined connectivity between various systems, and tailored medical intervention.

5.6.1 Personalized Healthcare: Empowering Patients for Better Outcomes

AI-powered blockchain technology holds immense promise in developing personalized healthcare solutions. By securely storing and sharing patient data on a decentralized ledger [14], healthcare providers can leverage AI algorithms to analyze vast datasets and deliver tailored treatments. Personalized healthcare not only enhances patient outcomes but also reduces costs by optimizing resource allocation and minimizing trial-and-error approaches.

5.6.2 Disease Surveillance and Prevention: Enhancing Public Health

The integration of AI and blockchain can revolutionize disease surveillance and prevention strategies. AI algorithms can analyze vast amounts of healthcare data, identifying patterns and trends that aid in predicting disease outbreaks and developing targeted interventions [15]. Through blockchain's immutable and transparent nature, public health agencies can securely track the spread of diseases, efficiently coordinate responses, and ensure data integrity.

5.6.3 Secure Data Management: Unlocking Interoperability and Collaboration

One of the significant challenges in healthcare is interoperability and secure data exchange between different organizations. Blockchain technology offers a decentralized and tamper-proof solution for managing health records. By employing smart contracts and access controls, patients can grant selective access to their data, enabling seamless collaboration among healthcare providers while maintaining data privacy and security.

5.6.4 Overcoming Challenges: Toward Widespread Adoption

While the potential of AI-powered blockchain technology in healthcare is promising, several challenges need to be addressed for its widespread adoption. These include regulatory considerations, scalability issues, standardization of data formats, and establishing trust among stakeholders. Collaboration between policymakers, technology developers, and healthcare professionals is vital in navigating these challenges and fostering a conducive environment for innovation.

The future of healthcare is poised for transformation with the integration of blockchain and AI technologies [16]. The potential applications discussed in this research paper, including personalized healthcare, disease surveillance, and secure data management, illustrate the profound impact this convergence can have on patient care and public health. Addressing the challenges associated with adoption is crucial for realizing the full potential of AI-powered blockchain technology in revolutionizing healthcare delivery and outcomes.

5.7 Conclusion

In closing, merging blockchain with AI into healthcare provides an enormous opening for improving patient outcomes while reducing costs through greater efficiency [17]. Blockchain technology offers secure storage while facilitating the sharing of medical data, while AI has unique abilities for managing large volumes of data analytics toward intelligible conclusions. In view of this approach, the incorporation of such technologies streamlines processes while enabling personalized care delivery by healthcare providers. Future integration can enhance applications in drug discovery methods ranging from remote patient monitoring techniques to predictive analytics-based solutions within our interactive health system framework that prioritizes patients through leveraging data usage optimally. Nevertheless, cognizance that certain challenges like maintaining privacy issues around data protection require collective stakeholder engagement for compliance on regulatory aspects arising from expansion efforts is critical. On balance, we look forward to seeing the trajectory as transformational trends persist toward greater implementation potentials throughout different areas including manipulation limitations possible by integrating blockchain with AI.

References

Journal Article

[1] Security of blockchain and AI-Empowered smart healthcare: Application-based analysis. (Link: https://www.mdpi.com/2076-3417/12/21/11039)

[2] Blockchain and artificial intelligence technology in e-Health. (Link: https://link.springer.com/article/10.1007/s11356-021-16223-0)

[3] Securing data with blockchain and AI (Link: https://ieeexplore.ieee.org/abstract/document/8733072)

[4] Blockchain technology in healthcare: Challenges and opportunities. (Link: https://www.tandfonline.com/doi/full/10.1080/20479700.2020.1843887)

[5] Applications of blockchain technology in medicine and healthcare: Challenges and future perspectives (Link: https://www.mdpi.com/2410-387x/3/1/3)

[6] Data sharing: Using blockchain and decentralized data technologies to unlock the potential of artificial intelligence: What can assisted

reproduction learn from other areas of medicine? (Link: https://www.sciencedirect.com/science/article/pii/S001502822032402X)

[7] Remote patient monitoring using artificial intelligence: Current state, applications, and challenges. (Link: https://wires.onlinelibrary.wiley.com/doi/full/10.1002/widm.1485)

[8] AI-powered blockchain technology for public health: A contemporary review, open challenges, and future research directions. (Link: https://www.mdpi.com/2227-9032/11/1/81)

[9] A systematic analysis on blockchain integration with healthcare domain: Scope and challenges (Link: https://ieeexplore.ieee.org/abstract/document/9449835)

[10] Blockchain and AI-empowered healthcare insurance fraud detection: An analysis, architecture, and future prospects. (Link: https://ieeexplore.ieee.org/abstract/document/9843995)

[11] Blockchain technology and artificial intelligence based decentralized access control model to enable secure interoperability for healthcare. (Link: https://www.mdpi.com/2071-1050/14/15/9471)

[12] Blockchain technology in the future of healthcare. (Link: https://www.sciencedirect.com/science/article/pii/S2352648321000453)

[13] Dimensions of artificial intelligence techniques, blockchain, and cyber security in the Internet of medical things: Opportunities, challenges, and future directions (Link: https://www.degruyter.com/document/doi/10.1515/jisys-2022-0267/html)

E-Book

[14] A critical analysis of the integration of blockchain and artificial intelligence for supply chain.

[15] Integration of artificial intelligence, blockchain, and wearable technology for chronic disease management: A new paradigm in smart healthcare.

[16] The blockchain technologies in healthcare: Prospects, obstacles, and future recommendations; lessons learned from digitalization.

[17] Blockchain–Internet of Things (IoT) enabled pharmaceutical supply chain for COVID-19.

Printed in the United States
by Baker & Taylor Publisher Services